CONSTRUCTION PRODUCTIVITY

A Practical Guide for Building and Electrical Contractors

Edited by

Eddy M. Rojas

Lic., M.S., and Ph.D. in Civil Engineering

M.A. in Economics

University of Washington

ISBN-13: 978-1-60427-000-6

Printed and bound in the U.S.A. Printed on acid-free paper
10987654321

Library of Congress Cataloging-in-Publication Data

Construction productivity : a practical guide for building and electrical contractors / Edited by Eddy M. Rojas.
p. cm.
Includes bibliographical references and index.
ISBN 978-1-60427-000-6 (hardcover : alk. paper)
1. Contractors--Handbooks, manuals, etc. I. Rojas, Eddy M.
TH12.C66 2008
690.68--dc22
2008029766

Direct all inquiries to J. Ross Publishing, Inc., 5765 N. Andrews Way, Fort Lauderdale, FL 33309.

Phone: (954) 727-9333
Fax: (561) 892-0700
Web: www.jrosspub.com

To all the men and women

who build our country

with their hands, minds, and hearts.

CONTENTS

PREFACE

Productivity is of paramount importance for the success of construction enterprises. This book is a collection of some of the best studies commissioned by ELECTRI International: The Foundation for Electrical Construction Inc. in the area of construction productivity. Studies commissioned by ELECTRI International are selected, coordinated, and monitored by some of the most progressive contractors in the construction industry and performed by outstanding scholars from some of the best universities in the United States. This combination of talents creates deliverables with valuable practical information that contractors can immediately apply and benefit from and that are based on sound methodological approaches.

This book is aimed at two distinct groups. First, it has become clear that most of the knowledge generated by the Foundation is universally applicable in the construction industry rather than exclusively to electrical construction. Therefore, in an effort to broadly disseminate its knowledge, the Foundation has decided to share some of its best studies for the benefit of the construction industry as a whole. Contractors from all trades will find value in the chapters included in this book, as they provide information that can be directly applied or easily adapted to any trade. Second, the Foundation has realized that educating the next generation of leaders is vital to ensuring the success of the construction industry. Therefore, this book also aims to reach out to students in civil engineering, construction management, building technology, and similar college programs by offering practical construction concepts to complement academic lectures.

Topics include how to measure labor productivity, establishing a field benchmarking program, dealing with stacking of trades, understanding and managing schedule acceleration and compression, understanding the relationships between absenteeism and turnover and labor productivity, negotiating loss of labor efficiency, developing field incentive programs, diminishing job stress in construction supervisors, and recommended practices for productivity enhancement.

A special feature of this book is the availability of free downloadable value-added resources in the form of practical, hands-on tools that can be applied to real-world situations. These resources are designed to enhance the learning experience as well as provide solutions to today's business challenges.

Finally, this book is part of a three-book series that disseminates ELECTRI International research studies. The other books in this series are *Construction Project Management* and *Construction Firm Management*.

ABOUT THE EDITOR

Dr. Eddy M. Rojas is a Professor in the Department of Construction Management at the University of Washington. He is also the Graduate Program Coordinator and the Executive Director of the Pacific Northwest Center for Construction Research and Education. He holds graduate degrees in civil engineering (M.S. and Ph.D.) and economics (M.A.) from the University of Colorado at Boulder and an undergraduate degree in civil engineering from the University of Costa Rica.

Throughout his academic career, Dr. Rojas has led numerous research studies in modeling, simulation, and visualization of construction engineering and management processes; engineering education; and construction economics. These studies have been sponsored by government agencies and private-sector organizations such as the National Science Foundation, the U.S. Department of Education, the U.S. Army, the Construction Industry Institute, the New Horizons Foundation, and ELECTRI International. Dr. Rojas has documented and disseminated the results and findings from his research efforts in numerous publications in technical refereed journals, technical conference proceedings, and technical reports, and in invited lectures and presentations at national and international seminars, symposia, and workshops.

Dr. Rojas is well known in both academic and professional circles not only through his research and publications but also by means of his professional activities, including his work as reviewer for the National Science Foundation, Specialty Editor for the *ASCE Journal of Construction Engineering and Management*, Chair and Technical Committee member in several congresses and conferences, reviewer for technical journals and conferences, and developer of the Virtual Community of Construction Scholars and Practitioners.

Free value-added materials available from the Download Resource Center at www.jrosspub.com

At J. Ross Publishing we are committed to providing today's professional with practical, hands-on tools that enhance the learning experience and give readers an opportunity to apply what they have learned. That is why we offer free ancillary materials available for download on this book and all participating Web Added Value™ publications. These online resources may include interactive versions of material that appears in the book or supplemental templates, worksheets, models, plans, case studies, proposals, spreadsheets, and assessment tools, among other things. Whenever you see the WAV™ symbol in any of our publications, it means bonus materials accompany the book and are available from the Web Added Value™ Download Resource Center at www.jrosspub.com.

Downloads for *Construction Productivity* include files on measuring productivity, field benchmarking procedures, and stacking.

1

MEASURING LABOR PRODUCTIVITY

Dr. Eddy M. Rojas, *University of Washington*

INTRODUCTION

The main goal of this chapter is to introduce practical and simple ways of measuring labor productivity in a contractor's organization at different levels of detail. Higher productivity levels allow contractors to simultaneously increase profitability, improve competitiveness, and pay higher wages to workers, while completing activities sooner. Once a contractor is capable of measuring labor productivity, the effect of any factor or combination of factors can be assessed by comparing labor productivity values before and after the factor was in place. For example, the impact of a new managerial approach in productivity can be determined by measuring labor productivity before and after the new approach is implemented.

In order to measure labor productivity, this chapter defines seven different levels at which the measurements are to be implemented. The highest level is the construction company. The second level divides company operations into two distinctive segments: administration and field operations. The third level is the project. The fourth level is the activity. The fifth level is the construction operation. The sixth level in this stratification is the process. Finally, the seventh and most basic level is the task.

Table 1.1 Productivity, Duration, and Cost

Item	Crew A	Crew B
Scope	200 fixtures	200 fixtures
Productivity	0.85 hrs/fixture	0.70 hrs/fixture
Time Required	170 labor-hours	140 labor-hours
Crew Size	4 workers	4 workers
Activity Duration	42.5 hrs	35.0 hrs
Time Savings		7.5 hrs
Wages	$35/hr	$38/hr
Indirect Cost	40%	40%
Labor Cost	$49.0/hr	$53.2/hr
Activity Labor Cost	$8,330	$7,448
Cost Savings		$882

WHY DOES LABOR PRODUCTIVITY MATTER?

Labor productivity is considered one of the best indicators of production efficiency. Higher productivity levels usually translate into superior profitability. A sustainable improvement in labor productivity is also associated with economic progress as it generates noninflationary increases in salaries and wages.

The main goal of this chapter is to introduce practical and simple ways of measuring labor productivity in a contractor's organization at different levels of detail. Once a contractor is capable of measuring labor productivity, the effect of any factor or combination of factors can be assessed by comparing labor productivity values before and after the factor was in place. For example, the impact of a new managerial approach in productivity can be determined by measuring labor productivity before and after the new approach is implemented. Furthermore, by calculating historical or typical productivity values, contractors can also identify projects under distress. Even though knowing that a project is under distress does not solve the situation, it allows contractors to start looking for the root causes of the problem and increases the chances of improving performance by implementing corrective actions early on.

The best way to understand the importance of labor productivity is through an example. Let's look at the installation of two-hundred 24-inch-long, two-lamp, industrial-grade fluorescent fixtures by two crews with different productivity levels.

Table 1.1 depicts the specifics of our case, where Crew B is more productive than Crew A. This higher productivity could be recognized by the contractor by offering higher hourly wages to Crew B workers. In Table 1.1, Crew B workers are paid $38/hr versus $35/hr for Crew A workers. However, even with this wage differential, the contractor still achieves savings of more than 10% in total labor costs ($882). These savings could (1) go directly to the contractor's bottom line to increase its profitability, (2) be fully transferred to the owner in an effort to improve the contractor's competitiveness, or (3) be allocated at will to increase both profits and competitiveness. Table 1.1 also shows that Crew B would complete the job almost one full day earlier than Crew A. Therefore, the higher productivity of Crew B allows the contractor to simultaneously increase its profitability, improve its competitiveness, and pay higher wages to its workers while completing the activity sooner. A truly win-win situation for all parties involved.

A LABOR PRODUCTIVITY MODEL FOR CONSTRUCTION

The term "labor productivity" is commonly used in a variety of contexts with different meanings. Formally, productivity is defined as the output produced per unit of time. All U.S. government data about labor productivity published by the Bureau of Labor Statistics are expressed in this manner. Generally, the U.S. government uses inflation-adjusted dollars to express output and an hour as the unit of time. Therefore, it is common to find government data where labor productivity is stated as, for example, $50/hr.

In the construction industry, because of tradition and practical reasons, productivity is usually expressed as the amount of time required to produce a unit of output, where the unit of output varies according to the circumstance. Therefore, it is common to find industry data, such as the National Electrical Contractors Association (NECA) Manual of Labor Hours, where labor productivity is stated as, for example, 1.06 hrs/fixture or 16 hrs/100 linear feet.

These two definitions of labor productivity are equivalent, as they both express efficiency. A labor productivity of 1.06 hrs/fixture can be translated into $/hr if one knows the total cost (material, labor, equipment, and indirect costs) of the fixture to the owner. For example, if we assume that the total cost of each fixture to the owner is $212, we obtain a productivity of $200/hr by dividing $212 by 1.06.

The traditional definition of 1.06 hrs/fixture is more meaningful in the field and more useful during the cost estimating process than the definition of $200/hr. However, the more formal definition of $200/hr is needed if we want to combine field productivity data from different activities to calculate the overall labor productivity for an entire project. Otherwise, how can we combine productivity data such as 1.06 hrs/fixture with 16 hrs/100 linear feet? This illustrates that no universal definition of labor productivity can be applied to all circumstances.

There are other issues that also complicate the process of measuring labor productivity. For example, should we exclude or include breaks, walking time, and

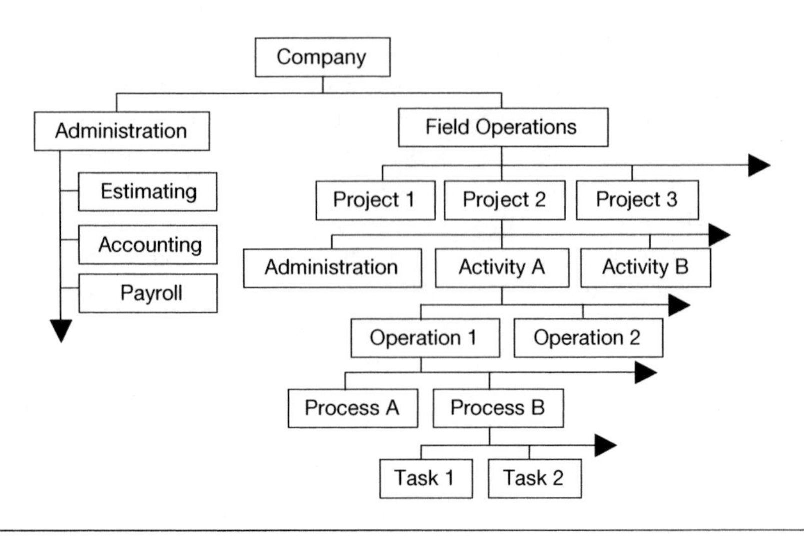

Figure 1.1 Construction firm stratification diagram

other so-called contributory activities when calculating the time required to install a fixture? Should we exclude or include lost time due to material unavailability, work stoppages, or interference from other trades? Should we focus on measuring productivity at the task level or at the process level when we can see how different activities interact with one another? Should we focus on measuring direct field productivity only or also the productivity of supervisors? Should we measure the productivity of administrative employees or only that of fieldworkers? Should we develop historical labor productivity data? And if so, how should we go about establishing baseline values? In order to answer these questions, it is necessary to define a labor productivity model for construction that can be used as a guideline to facilitate and simplify the process of measuring labor productivity.

Figure 1.1 shows a construction firm stratification diagram. At the top of this stratification we have the construction company, which can be defined as an organization whose primary emphasis is to build a facility or some systems in a facility in order to satisfy customer's expectations by providing quality at a reasonable cost while complying with industry and government standards regarding safety and environmental issues. At the company level, the main focus is on strategic business decisions such as market selection, definition of products and services, definition of growth and profitability targets, selection of personnel, and establishment of a succession plan among others.

The second level divides company operations into two distinctive segments: administration and field operations. Administration is commonly referred as home office overhead, and it is considered a necessary but non-value-added activity from the owner's viewpoint.

The third level is the project. At the project level the main focus in on scope definition, identification of project requirements and construction activities, development of contractual arrangements, and planning and control of cost, time, and resources.

The fourth level is the activity. A construction activity is a physical segment of a project that consumes specific resources and requires a given amount of time for completion. Projects are broken down into activities to facilitate scheduling, as well as time and cost control. At the activity level the main focus is on early and late start and completion dates, activity float, percentage of completion, and resource consumption.

The fifth level is the construction operation. A construction operation is not a physical but a technological segment of a project. Therefore, a construction operation can span several activities. In other words, the same construction operation could be part of more than one activity. A construction operation is also cyclical or repetitive. At the operation level the main focus is on construction methods and technologies.

The sixth level in this stratification is the process. A process is a segment of a construction operation that uses the same resources. In simple operations, when only one process exists, process and operation become synonymous. The main focus at the process level is on technological sequencing and logical collection of tasks.

Finally, the seventh and most basic level is the task. A task is an elemental work assignment. The main focus at the task level is on intrinsic knowledge and skills of crew members.

The best way to understand this stratification or hierarchy of construction and the definitions provided above is to analyze an example. Therefore, let's look at an electrical contractor with 200 employees. This contractor may be working on several projects simultaneously. We can focus on one of those projects: the electrical installations for a new semiconductor plant. For managerial purposes, this project is broken down into a variety of activities. Let's examine one of these activities closely: Switchboard Installation Building A. This activity identifies a segment of the project in a particular location, building A. Other switchboards will likely be installed in this project, and this is why identifying the particular location is important. This activity will probably be part of the critical path and as such is crucial for the on-time completion of the project. Different accounts will also be created to keep track of all expenses in this activity. For typical project control purposes, this is the lowest level of detail that a construction manager would deal with. However, from a technological perspective we can go further. Figure 1.2 shows the stratification of this activity.

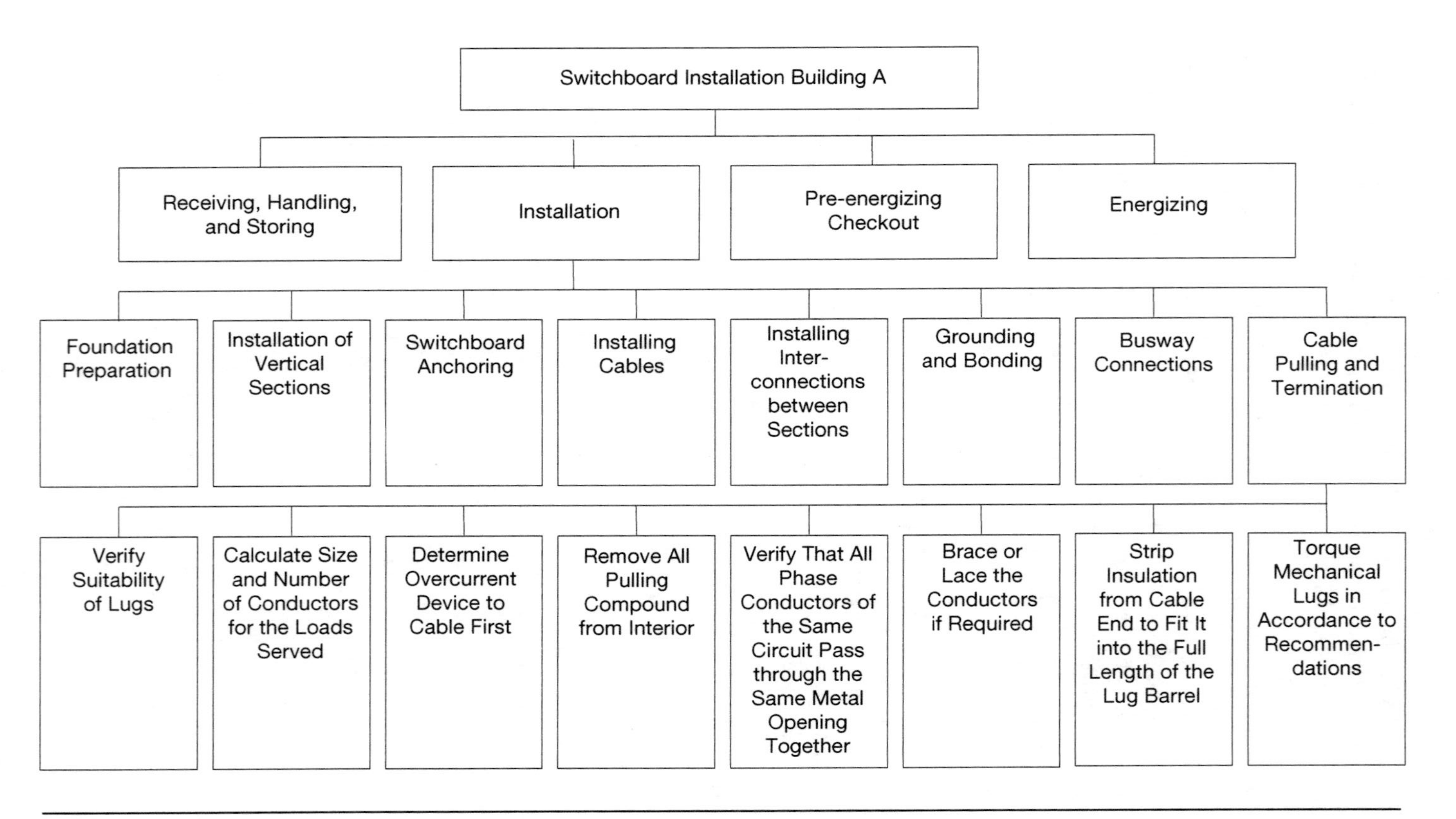

Figure 1.2 Stratification diagram for the "Switchboard Installation Building A" activity.

The activity "Switchboard Installation Building A" includes four independent construction operations:

1. Receiving, handling, and storing
2. Installation
3. Pre-energizing checkout
4. Energizing

These are four separate operations because they use different methods, technologies and labor skills.

Let's examine one of these operations in greater detail: Installation. The installation of this switchboard will require eight processes:

1. Foundation preparation
2. Installation of vertical sections
3. Switchboard anchoring
4. Installing cables
5. Installing interconnections between sections
6. Grounding and bonding
7. Busway connections
8. Cable pulling and termination

Let's now look at the last process, Cable pulling and termination, to identify the individual tasks that must be performed in order to complete this process:

1. Verify suitability of lugs.
2. Calculate size and number of conductors for the loads served.
3. Determine overcurrent device to cable first.
4. Remove all pulling compound from interior.
5. Verify that all phase conductors of the same circuit pass through the same metal opening together.
6. Brace or lace the conductors if required.
7. Strip insulation from cable end to fit it into the full length of the lug barrel.
8. Torque mechanical lugs in accordance with recommendations.

As we can see from this example, the stratification or hierarchy of construction defined in this section is a useful tool to analyze and break down the construction process at different levels of detail.

MEASURING LABOR PRODUCTIVITY AT DIFFERENT LEVELS

Before we begin analyzing labor productivity at different levels of construction stratification, we must remind ourselves that construction companies are systems. A system is defined as a collection of interrelated components. Therefore, in order to enhance the efficiency of a construction company, we must work at it from a system's point of view, which involves looking at the overall picture. The process of optimizing an individual component in a system may not produce the desired results if that component is not the one causing the problem or if other necessary components are not also improved concurrently. For example, increasing the productivity of workers may not achieve the desired result of improving efficiency if the procurement processes are not modified to also provide tools, equipment, and materials at the higher rates required. One can imagine a very productive crew having to wait an hour for required materials and not achieving the level of production desired. Therefore, the analysis that follows, where we look at measuring labor productivity at different levels, is not intended to convey the impression that labor productivity can be improved at each level in isolation and without regard to what is happening at the other levels. On the contrary, the purpose of looking at different levels is to uncover information that would help us diagnose problems and apply solutions to the system. For example, significant problems with availability of materials when measuring the productivity of one activity should be interpreted as a symptom of a potentially bigger administrative problem that may require the redesign of the company-wide procurement system, not only the procurement tasks performed for that particular activity.

Company Productivity

The intent of measuring labor productivity at the company level is to determine the overall performance of a contracting organization. A basic definition of company level productivity is as follows:

$$P_{\text{COMPANY-CURRENT}} = \frac{\text{TAVA}}{\text{TAL}} \tag{1.1}$$

where:

$P_{\text{COMPANY-CURRENT}}$:	Productivity at the company level (not adjusted for wage inflation)
TAVA:	Total accrued value added
TAL:	Total accrued labor hours

TAVA is calculated as:

TAVA = Acrrued [Income–Materials–Subcontracts–Owner paid overtime premiums] (1.2)

The concept of TAVA is paramount in the definition of labor productivity used in this chapter. At a first glance, we may want to include all revenue generated by a contracting organization when calculating labor productivity. However, doing so may prevent us from comparing "apples to apples." Contractors use different financial and organizational arrangements when delivering projects. In some instances, a contractor may buy all materials and self-perform all activities. In other instances, a contractor may install materials provided and bought by the owner and subcontract some installations.

A project where no materials are provided by the owner would show significantly higher revenues than if materials were provided. However, the number of labor hours required to install those materials is the same regardless of who buys the materials. If we do not correct our labor productivity calculation to consider this situation, then the same project would show different labor productivity values depending on who is buying the materials.

A project where all installations are self-performed would require more labor hours than if some of the installations were subcontracted. However, the revenue generated is the same in both instances. If we do not correct our labor productivity calculation to consider this situation, then the same project would show different labor productivity values depending on how much of its scope is subcontracted.

Overtime pay must be considered in a slightly different light. If overtime paid is the result of a contractor's decision to expedite the completion of the project and the labor premium paid is born exclusively by the contractor, then this practice has no bearing in our calculation of labor productivity, as it does not directly affect either the revenue or the number of hours worked. However, if the overtime is the result of an owner directive where the owner is compensating the contractor for the additional cost, then this will introduce a distortion in our calculation of labor productivity, as the revenue will be higher for the same number of hours.

In order to eliminate all of these distortions, we use the concept of value added. Value added can be defined as the additional value given to a set of materials when they are installed in a project. In other words, a fully assembled electrical system has more value to the owner than the sum of its parts (adding the material value of each component). In addition, the value added by a contractor only includes the self-performed installations. You do not add value to a project by the work performed by a subcontractor, since it is the subcontractor who is actually adding that value. Furthermore, value is not added to a project by having the owner pay an overtime premium.

TAVA includes the term "accrued" in its description because it consists of all income earned by a company for the sale of goods and services in a specific period of time minus material purchases, subcontracts, and owner-paid overtime premiums. Notice that the word "earned" is used rather than "collected" because accrued income includes receivables that a company may have for goods and services already rendered and excludes payments made during the period for revenues earned during the previous period. Accountants should be able to compute TAVA for a contracting organization.

The second value required to compute labor productivity at the company level is total accrued labor hours, or TAL. TAL includes all labor hours (fieldworkers and home office personnel) that the company has benefited from during a specific period of time whether or not payment has been rendered. Therefore, hours for salaries and wages payables must also be included in this calculation, and wages and salaries paid in this period for work performed in the previous period must be excluded. Once again, accountants should be able to compute TAL for a contracting organization.

Field labor hours can be determined from field records, or in the absence of detailed information, they can be approximated. An approximation of the total number of field labor hours can be calculated by dividing total accrued wages of fieldworkers by the average hourly wage. Home office personnel hours can be determined by calculating the number of full-time-equivalent (FTE) positions and multiplying this by the number of working hours for a specific period of time. For example, one can use 500 hours per quarter or 2,000 hours per year.

The simple measure of productivity at the company level introduced in Equation 1.1 presents a problem when comparing productivity over time as it does not consider the effect of wage inflation. Therefore, a better formulation is as follows:

$$P_{COMPANY} = \left(\frac{TAVA}{TAL}\right)\left(\frac{WBY}{WCY}\right) \quad (1.3)$$

where:

$P_{COMPANY}$:	Productivity at the company level
TAVA:	Total accrued value added
TAL:	Total accrued labor hours
WBY:	Average wages base year
WCY:	Average wages current year

The factor (WBY/WCY) shown in Equation 1.3 is used to convert current dollars to dollars from a selected base year. Any previous year can be used as a base year. This adjustment for wage inflation allows the comparison of company-level labor productivity values across time, which in turn facilitates the identification of trends.

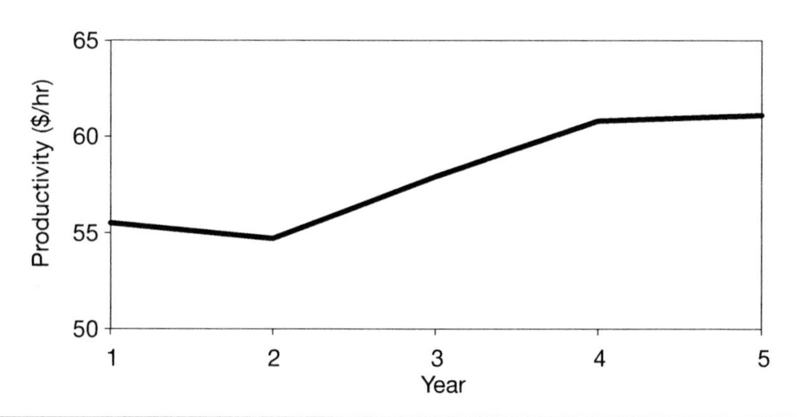

Figure 1.3 $P_{COMPANY}$ (Equation 1.3).

Company-level labor productivity can also be expressed in alternative ways. For example, one may want to express labor productivity as the number of hours required to generate a given amount of value added. Equation 1.4 shows the formula required to implement this calculation.

$$P_{COMPANY\text{-}HOURS} = \left(\frac{TAVA}{TAL}\right)\left(\frac{WCY}{WBY}\right)(VAM) \tag{1.4}$$

where:

$P_{COMPANY\text{-}HOURS}$:	Productivity at the company level (in hours per given revenue)
TAL:	Total accrued labor hours
TAVA:	Total accrued value added
WCY:	Average wages current year
WBY:	Average wages base year
VAM:	Value-added multiplier

The VAM term in Equation 1.4 refers to the given amount of value added used as the basis for the calculation. For example, if $10,000 is used for the VAM, then Equation 1.4 would calculate the number of labor hours required to generate $10,000 of value added. Therefore, while Equation 1.3 provides values in dollars per hour, Equation 1.4 provides values in hours per given amount.

Contractors may also want to create their own company productivity indexes. This can be accomplished by normalizing the value of company-level labor productivity for a given year to be equal to 100. This is the function of the IF term shown on Equation 1.5. Then all other years will be expressed in relation to this base year.

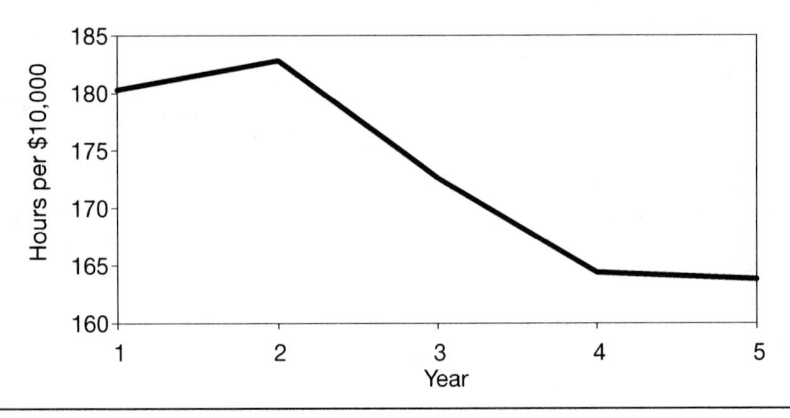

Figure 1.4 $P_{COMPANY\text{-}HOURS}$ (Equation 1.4).

$$\text{P-Index}_{COMPANY} = \left(\frac{TAVA}{TAL}\right)\left(\frac{WBY}{WCY}\right)(IF) \tag{1.5}$$

where:

$\text{P-Index}_{COMPANY}$:	Productivity index at the company level
TAVA:	Total accrued value added
TAL:	Total accrued labor hours
WBY:	Average wages base year
WCY:	Average wages current year
IF:	Index factor

The value of IF can be calculated by using a simple formula:

$$IF = \frac{100}{P_{COMPANY} \text{ for the base year}} \tag{1.6}$$

By normalizing labor productivity with the use of a company-specific index, the data can be easily shared with field office personnel without the fear of divulging highly sensitive proprietary data.

Example 1.1: A small contractor wants to calculate its company-level labor productivity for the last five years in order to evaluate its overall performance. As shown on Table 1.2, this contractor did not have complete records for hours worked for the entire five-year period. Therefore, for years 1 and 2, the number of field labor hours is approximated by dividing total wages by the average field wage paid. After performing the calculations, one can observe that this company's labor productivity, according to Equation 1.3 (Figure 1.3), decreased from

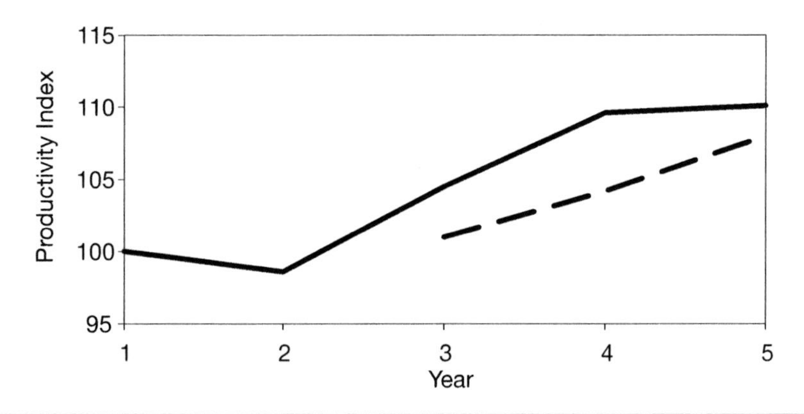

Figure 1.5 P-Index $_{COMPANY}$ (Equation 1.5).

$55.5/hr for the year 1 to $54.7/hr for year 2. This small variation in labor productivity values is normal and should not raise any red flags. However, for the last three years an upward trend is observed, as the company's productivity increased to $57.9/hr, $60.8/hr, and $61.1/hr in years 3, 4, and 5, respectively. This is good news for the contractor!

When productivity is calculated with Equation 1.4, an inverse relationship is evident: Higher productivity as measured in $/hr is equivalent to fewer hours required to generate a given amount of value added—$10,000 in this example. Therefore, Equation 1.4 (Figure 1.4) provides a different but equivalent way of analyzing the data as compared with Equation 1.3. Finally, Equation 1.5 normalizes the data calculated by Equation 1.3 to create a company-based productivity index using the productivity of year 1 as the base (productivity of year 1 = 100). As shown by the corresponding graphs, the line generated with Equation 1.3 (Figure 1.3) and the line generated by Equation 1.5 (Figure 1.5) are equivalent, since Equation 1.5 is only a transformation of scale of Equation 1.3.

In the calculation of productivity values, small variations can be expected from year to year. This is absolutely normal and should not be a source of concern. Therefore, it may be necessary to smooth out the data in order to facilitate the process of identifying trends. Table 1.2 shows the calculation of a three-year moving average for the values obtained from Equation 1.5. A three-year moving average is calculated by adding the productivity value for a given year to the productivity values from the preceding two years and dividing the result by 3. Since three years of data are needed, the three-year moving average can only be calculated for years 3, 4, and 5. Figure 1.5 shows the three-year moving average in dotted lines. An upward trend is a positive finding for a contractor, and productivity values above a moving average represent an improvement in productivity beyond the company's historical trend, as is the case in Figure 1.5. On the other

Table 1.2 Five Years of Productivity Data for a Small Contractor

	Year				
	1	2	3	4	5
Known Variables:					
Accrued Income	13,250,000	14,000,000	15,800,000	20,500,000	25,200,000
Accrued Value of Materials	5,300,000	6,300,000	5,530,000	8,200,000	10,584,000
Accrued Value of Subcontracts	550,000	245,000	350,000	750,000	425,000
Field Labor Hours			139,500	150,000	178,000
Field Wages	5,400,000	5,500,000			
Average Field Wage	46	48	50	52	54
Home Office FTE	8	8	9	9	10
Calculated Variables:					
Total Accrued Labor Hours	133,391	130,583	157,500	168,000	198,000
$P_{COMPANY}$ (Equation 1.3)	55.5	54.7	57.9	60.8	61.1
Value Added Multiplier	10,000				
$P_{COMPANY\text{-}HOURS}$ (Equation 1.4)	180.3	182.8	172.6	164.4	163.8
P-Index $_{COMPANY}$ (Equation 1.5)	100.0	98.6	104.5	109.6	110.1
P-Index $_{COMPANY}$ 3-year Moving Average			101.0	104.2	108.0

hand, productivity values below a moving average show performance below historical trends. A productivity value below a moving average, though, may still represent an improvement in productivity from the previous year. This would be the case for a contractor who has been improving productivity at a given rate, failed to maintain the same pace of improvement for the current year, yet increased productivity nonetheless.

Field Operations Productivity

The intent of measuring labor productivity at the field level is to determine the level of performance of a contracting organization in the field, independent from the home office operations. Productivity at the field operation level is given by the following:

$$P_{FIELD} = \left(\frac{TAVA}{AFL}\right)\left(\frac{WBY}{WCY}\right) \tag{1.7}$$

where:

P_{FIELD}:	Productivity at the field level
TAVA:	Total accrued value added
AFL:	Accrued field labor hours
WBY:	Average wages base year
WCY:	Average wages current year

The only difference between Equation 1.7 and Equation 1.3 is the number of labor hours considered. While Equation 1.3 considers all labor hours (field plus home office), Equation 1.7 only includes field labor hours.

Productivity of field operations can also be expressed as the number of hours required to generate a given amount of value added (value-added multiplier). Equation 1.8 shows the formula required to perform this computation.

$$P_{FIELD\text{-}HOURS} = \left(\frac{AFL}{TAVA}\right)\left(\frac{WCY}{WBY}\right)(VAM) \tag{1.8}$$

where:

$P_{FIELD\text{-}HOURS}$:	Productivity at the field level (in hours per given revenue)
AFL:	Accrued field labor hours
TAVA:	Total accrued value added
WCY:	Average wages current year
WBY:	Average wages base year
VAM:	Value-added multiplier

Table 1.3 Five Years of Field Productivity Calculations for a Small Contractor

	Year				
	1	2	3	4	5
Calculated Variables:					
Accrued Field Labor Hours	117,391	114,583	139,500	150,000	178,000
P_{FIELD} (Equation 1.7)	63.0	62.4	65.4	68.1	67.9
Value Added Multiplier	10,000				
$P_{FIELD\text{-}HOURS}$ (Equation 1.8)	158.6	160.4	152.9	146.8	147.2
P-Index $_{FIELD}$ (Equation 1.9)	100.0	98.9	103.8	108.1	107.7
P-Index $_{FIELD}$ 3-year Moving Average			100.9	103.6	106.5

Once again, Equation 1.8 is analogous to Equation 1.4, and the only difference is the number of labor hours included in the calculation. Finally, contractors may also want to create their own productivity index for field operations, using an analogous formula to the one shown on Equation 1.5:

$$\text{P-Index}_{\text{FIELD}} = \left(\frac{\text{TAVA}}{\text{AFL}}\right)\left(\frac{\text{WBY}}{\text{WCY}}\right)(\text{IF}) \tag{1.9}$$

where:

P-Index $_{FIELD}$:	Productivity index at the field level
TAVA:	Total accrued value added
AFL:	Accrued field labor hours
WBY:	Average wages base year
WCY:	Average wages current year
IF:	Index factor

Example 1.2: The small contractor from Example 1.1 now wants to calculate its five-year productivity values at the field operations level. The required data for this analysis is located in Table 1.2. After performing the necessary operations, the results are shown on Table 1.3 and Figures 1.6, 1.7, and 1.8.

Not surprisingly, the field labor productivity of our small contractor follows the same trend as the company's labor productivity. For trade contractors who self-performed most or all of their work, field labor hours will represent the majority of labor hours and will dictate the trends at both the company and the field level. General contractors or trade contractors who subcontract a significant

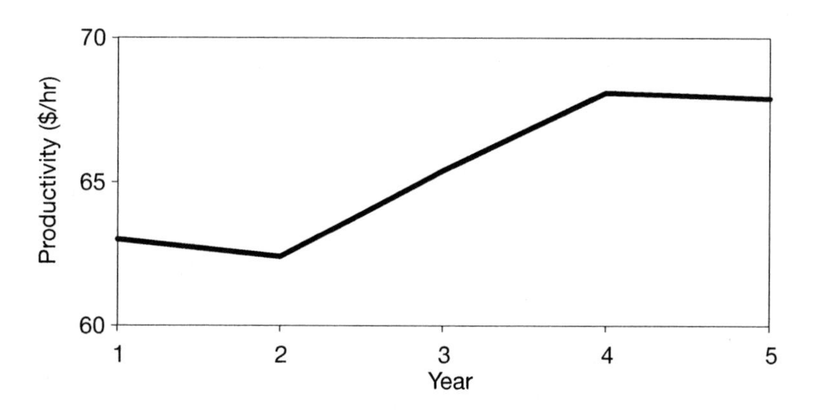

Figure 1.6 P_{FIELD} (Equation 1.7).

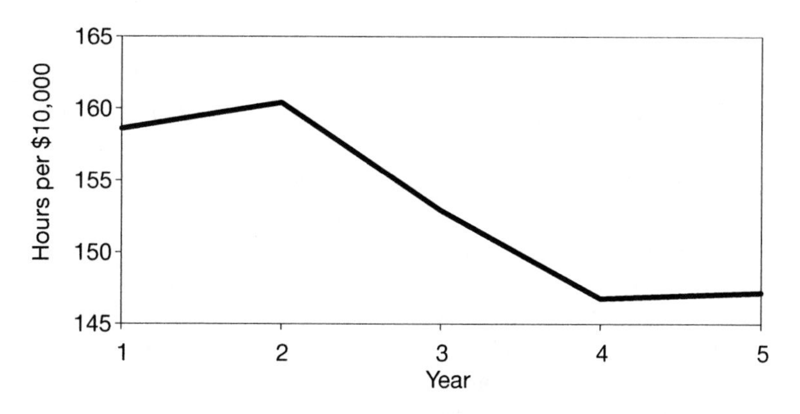

Figure 1.7 $P_{FIELD\text{-}HOUR}$ (Equation 1.8).

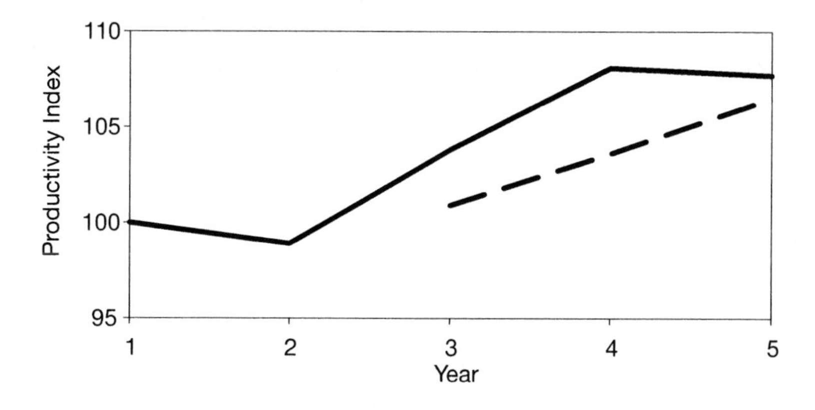

Figure 1.8 P-Index $_{FIELD}$ (Equation 1.9).

amount of work may experience a greater divergence between labor productivity trends at the company and the field levels.

Going back to our example, it may be difficult to compare the labor productivity at the company level to the labor productivity at the field level. The best way to do so is to look at the productivity indexes calculated for both cases (same base year required). As we can see by reviewing the data in Table 1.2 and Table 1.3, the productivity index at the company level is slightly higher than the productivity index at the field level. This implies that home office personnel have increased their labor productivity at a higher rate than fieldworkers over the past four years. This is a typical result, as it is easier to increase the productivity of home office personal than the productivity of fieldworkers. After all, any contractor would argue that it is possible to handle one more project per year with the same home office personnel, but not with the same number of fieldworkers (hours).

Project Productivity

The intent of measuring labor productivity at the project level is to determine the level of performance of a contracting organization by project. This is important, as patterns or correlations may be uncovered. For example, an electrical contractor may realize that its productivity tends to be 10% lower when working with a given general contractor (GC). In light of this information, the electrical contractor may choose not to work with that particular GC in the future, or alternatively, the contractor may choose to increase the amount budgeted for labor in bids by 10% when sending them to this GC. Contractors may also find patterns related to project managers. A project manager who consistently achieves the lowest labor productivity values within the company may have to be let go in order to enhance the company's efficiency. Furthermore, contractors may also be able to find patterns related to market segments. Contractors who work in several markets may not be as efficient in all of them. Knowing the areas in which a contractor has a competitive advantage (its higher labor productivity values) may prove valuable when making strategic decisions about areas of expansion for the company.

Productivity at the project level is given by the following:

$$P_{PROJECT} = \left(\frac{TAVAP}{ALH}\right)\left(\frac{WBY}{WCY}\right) \quad (1.10)$$

where:

$P_{PROJECT}$: Productivity at the project level
TAVAP: Total accrued value added of project
ALH: Accrued labor hours for the project
WBY: Average wages base year
WCY: Average wages current year

The TAVAP term includes all value added for work invoiced or to be invoiced in a project, including change orders. The ALH term includes all direct labor consumed by the project, accounting for both craftspeople and supervisory employees. However, only those supervisory employees who work full-time for the project should be included. Part-time supervisory employees, such as project managers who oversee multiple projects, are considered part of the home office personnel. Notice that average wages are used as wage inflation indexes so that data from projects performed in different years can be compared.

Finally, contractors may also want to create their own productivity index for projects, using an analogous formula to the one shown on Equation 1.5:

$$\text{P-Index}_{\text{PROJECT}} = \left(\frac{\text{TAVAP}}{\text{ALH}}\right)\left(\frac{\text{WBY}}{\text{WCY}}\right)(\text{IF}) \qquad (1.11)$$

where:

$\text{P-Index}_{\text{PROJECT}}$:	Productivity index at the project level
TAVAP:	Total accrued value added of project
ALH:	Accrued labor hours for the project
WBY:	Average wages base year
WCY:	Average wages current year
IF:	Index factor

Contractors may select any project as the base project for the index calculation. Once again, by normalizing labor productivity with the use of a project-specific index, the data can be easily shared with field office personnel without the fear of divulging highly sensitive proprietary data.

Example 1.3: A contractor wants to calculate the labor productivity values for three of his projects. The required data for this analysis, as well as the results of the calculations, are shown on Table 1.4.

Project 1 and Project 2 show similar values for labor productivity, at around \$64/hr. Project 3, on the other hand, shows a significantly higher value at \$72.4/hr. This difference may be explainable if Project 3 belongs to a different market than Projects 1 and 2 or because it has some unique features. However, if these three projects belong to the same market and are similar in nature, then Project 3 represents an outstanding performance when compared to the other two projects. If this is the case, then the contractor should investigate further and verify the validity of the data. If the data is correct, this would represent a significant finding for the company. Further analysis of Project 3 would be warranted to learn what possible factors could have contributed to this outstanding performance and to evaluate the possibility of transferring lessons learned to future projects. Also notice

Table 1.4 Labor Productivity for Three Projects

	Project		
	1	2	3
Known Variables:			
Project Value Added	1,650,000	2,475,000	3,712,500
Labor Hours	25,756	38,345	49,897
Average Wages	37	37	38
Calculated Variables:			
$P_{PROJECT}$ (Equation 1.10)	64.1	64.5	72.4
P-Index $_{PROJECT}$ (Equation 1.11)	100.0	100.8	113.1

that Projects 1 and 2 have the same average wages, while Project 3 has a different one. This may indicate that Projects 1 and 2 were built in the same year, but Project 3 was built in a different year. Alternatively, this may also indicate that Project 3 was built in a different jurisdiction than Projects 1 and 2. Therefore, the adjustment that we have been using to account for wage inflation can also be used to compare projects built during the same year at different jurisdictions (locations with different average wage values).

Activity Productivity

The intent of measuring labor productivity at the activity level is to determine the level of performance at a greater detail than possible when looking at the company, the field, or even the project level, but still at a high enough level of detail to identify problems with coordination issues.

Productivity at the activity level is given by:

$$P_{ACTIVITY} = \left(\frac{TAVAA}{ALHA}\right)\left(\frac{WBY}{WCY}\right) \tag{1.12}$$

where:

$P_{ACTIVITY}$: Productivity at the activity level
TAVAA: Total accrued value added of the activity
ALHA: Accrued labor hours for the activity
WBY: Average wages base year
WCY: Average wages current year

The TAVAA term includes all value added for work invoiced or to be invoiced for the activity, including change orders. The ALHA term includes only direct labor

Table 1.5 Labor Productivity for Two Activities

	Activity	
	1	2
Known Variables:		
Activity Value Added	175,000	262,500
Labor Hours	2,862	3,486
Average Wages	37	37
Calculated Variables:		
$P_{ACTIVITY}$ (Equation 1.12)	61.2	75.3

consumed by the activity. No supervisory employees should be included in this calculation. Notice that average wage rates are used so that data from similar activities performed in different years (or different jurisdictions) can be compared.

Example 1.4: A contractor wants to calculate the labor productivity values for two critical activities. The required data for this analysis, as well as the results of the calculations, are shown in Table 1.5.

The calculation of the labor productivity values for the two activities shown in Table 1.5 follows the same basic procedure that we have used previously. A productivity index for activities can also be calculated, if desired, following an analogous procedure to the one followed at the company, field, and project levels. However, there is a significant difference in how the "known" variables are obtained. Calculating labor productivity at the company, field, or project level can be accomplished by processing data that most accounting departments already have as a result of their regular operations. However, calculating labor productivity at the activity level requires information from the field that is usually not captured during the performance of regular company business.

Figuring out the value added of the activity is not as simple as getting the value added for an entire project, as we did when calculating project productivity. First, one must clearly define what the scope of the activity is. Second, most likely the procurement process does not keep track of the amount or cost of the materials on an activity-by-activity basis. Third, the appropriate indirect cost items must be selected as part of the value of the activity. These indirect cost items include profits, contingency, financial charges, bonding and insurance, equipment, and supervision, among others. Perhaps the simplest procedure to accomplish this task is to develop a detailed estimate of the activity at hand as if the work were to be performed via a change order, and use this amount as the total value of the activity from which materials, subcontracts, and owner-paid overtime premiums are to be subtracted.

Figuring out labor hours is not simple either, as workers usually work on several activities during the same week or even the same day. Properly allocating labor hours to an activity may require additional record keeping in the form of timesheets for each employee who works on the activity.

Because additional information must be collected, it is impractical to calculate labor productivity for all activities in a given project. Furthermore, not all activities are equally important. Activities that are repetitive, belong to the critical path, or involve a great deal of uncertainty regarding labor consumption are prime candidates for measuring productivity. An electrical contractor who wishes to measure the productivity of key construction activities in a project should begin by inviting field personnel to actively participate in the process. A simple data collection system consisting of forms to be filled out by electricians and foremen should be designed to ensure minimum disruption of the activity and reduce the additional effort required to collect the data.

It is imperative to communicate to those in the field that the objective of measuring activity productivity is to evaluate the efficiency of the activity and not the performance of the workers. Studies have shown that the majority of the unproductive time for construction workers is usually caused by problems with the process rather than by inefficiencies or laziness on the part of workers. Examples of this include lack of proper directions, tools, or materials; requests to go to workstations that are not ready for them or that are already occupied; requests for rework because of changes; and requests to work in unreasonable sequences or using inefficient methods.

In addition to obtaining quantitative information about construction activities, contractors should take advantage of the participation of field personnel to capture some qualitative data that can be as valuable as the numerical results. We have to keep in mind that the individuals who are actually performing the work are in one of the best positions to provide feedback about what is and what is not working properly. Surveys and interviews should be performed to gather this qualitative data, since traditional channels of communication, which require the information to go through several levels, are usually unreliable because people at each level tend to filter pieces of information for a variety of reasons. Therefore, significant problems may never be uncovered. Questionnaires and interviews, when used properly, are great tools to uncover problems and gather the opinions of workers in an unbiased fashion.

Questionnaires must be carefully designed to ensure that they will uncover problems. There are many ways to ask a question, from multiple-choice to open-ended written statements. Multiple-choice questions are appropriate when there are only a few options available. On the other hand, multiple-choice questions must be avoided when no one knows exactly how many different answers there may be for a question. Open-ended questions are appropriate when one does not want to limit the respondents to a preselected number of options. However, most people do not like to spend a lot of time answering open-ended questions. A proper combination of open-ended with other type of questions may be best. A "yes or no" question is another technique for gathering information. These questions force respondents to take positions, since they cannot stay in the middle of an issue. For example, they can either say that the instructions provided are the best for the job or not, but they cannot state that they are just OK. Another option is to ask people to grade the level of a problem or the frequency of occurrence of specific situations. For example, for a list of issues one may ask if they are "very important," "important," or "not really important," or one may ask if they occur "very frequently," "frequently," "sometimes," or "almost never."

Contractors should make clear that answering the questionnaire or participating in an interview is voluntary and that anyone who feels uncomfortable may be excused. Common topics covered in questionnaires and interviews include (1) availability and appropriateness of materials, tools, and equipment; (2) suitability of instructions and other information; (3) availability and suitability of safety equipment; (4) sequencing of tasks; (5) frequency and reasons for rework; (6) frequency and reasons for interference with other crafts or activities; and (7) problems with overmanning, overtime, turnover, and absenteeism.

Finally, it is imperative to keep in mind that information obtained from interviews or questionnaires is basically useless unless lessons learned are extracted by carefully studying the answers given and determining the root causes of problems. The answers will only reveal symptoms of problems rather than the problems themselves. Once the root causes of the problems are identified (i.e., lessons are learned), then corrective actions must be implemented company-wide to improve the processes involved.

Operation and Process Productivity

One of the disadvantages of calculating productivity at the activity level is that the units of measurement are dollars per hour. This unit of measure is necessary because an activity, by definition, involves several operations that have their own units of measurement. The only way to combine productivities of operations measured in different units (i.e., fixtures/hr vs. linear feet/hr) is to use dollars per hour. However, dollars per hour is not a very amenable unit to evaluate field productivity. Therefore, in many cases, contractors may want to collect only qualitative information at the activity level through questionnaires and interviews and leave the measurement of productivity at the operation or the process level. This section deals with measuring the productivity of operations and processes because in many cases an operation has only one process and, therefore, these two concepts become synonymous. Furthermore, even when an operation has several processes, the techniques used to measure the productivity of operations are basically the same as those used to determine the productivity of processes.

A variety of tools are available to measure labor productivity at the operation or process level, among them: (1) time studies, (2) photographic and video studies, and (3) activity sampling.

Time Studies

Time studies are used to collect information about the duration of the different tasks that make up a construction operation or process. Therefore, these studies are task-based rather than worker-based. Data-gathering forms must be prepared in advance to collect the information, and they will be specific for each construction operation or process. Traditional time studies require a significant amount of resources, as one observer needs to be assigned to each worker. Traditional time studies have some limitations due to their nature. For example, the observer must make a determination about the instantaneous point in time when a task finishes and the next one begins. For some construction operations or processes, such a point in time may be more obvious than for others. This is not usually a problem when the same person is collecting all data. However, a major concern may arise when different observers are collecting data on the same operation or process and the data is to be combined later. Additionally, most observers usually do not collect explanatory data that could shed some light on why a particular task or set of tasks is taking significantly more time than expected. This happens because of the large amount of data that is often collected and the time and physical effort required.

A modified version of time studies can be applied, however, if workers are recruited to collect their own data. Obviously, the main reason behind having observers collect the data is lack of trust of workers. Once again, it is imperative to communicate to those in the field that the objective of measuring productivity is to evaluate the efficiency of the operation and not the performance of the workers. Worker trust and cooperation is indispensable if modified time studies are to be performed. Furthermore, workers can add information of great value to their time annotations by including notes and comments that help contextualize the information. Notes regarding why their productivity was negatively or positively affected during a particular operation can be invaluable to management. Forms must be designed to allow workers to collect their own data. These forms should capture amounts of materials or equipment installed, time invested in every task, and contextual information to explain deviation from expected performance. Management can then combine the data from these forms and calculate productivity values in meaningful units such as hr/fixtures or hr/linear feet.

Photographic and Video Studies

Photographic and video studies can be very useful tools in studying construction operations and processes. They can capture at least the same information as time studies without requiring the expense of several people taking notes.

Photographic studies refer to the use of a camera or video camera to capture time-lapse pictures of a construction operation. Time-lapse refers to the process of taking still pictures at selected intervals. In general, the interval can vary from 1 to 10 seconds. The shorter the interval, of course, the greater the amount of detail that can be gathered. Intervals of 1 second are usually used when you want to keep track of hand movements. Intervals of 4 seconds are common when you want to study the overall body or equipment movements. Intervals greater than 4 seconds are used only as a general reference on how the operation is proceeding. Video studies refer to the use of a video camera to capture real-time video of a construction operation. Digital video cameras and video processing software can allow real-time video to be recorded and time-lapse pictures from the same event to be extracted. Following are some of the advantages of photographic and video studies:

- A permanent record of the operation can be created that can be studied from several perspectives by a variety of individuals.
- By using time-lapse photography, operations, or processes that take hours can be studied in minutes.
- The video can be shown to craftspeople, foremen, and supervisors to gather their opinions regarding improvement in productivity, quality, and safety.
- Inefficiencies and unbalances such as interferences among crew members, lack of materials or tools, and changes in procedures, among others, can also be uncovered.
- The detailed records provided by these studies can help resolve disputes in the project.
- The photographic/video record can provide excellent support for or refutation of claims.
- They can be excellent training and teaching aids.
- They can be used to capture and support lessons learned.

Following are some issues you should keep in mind when performing photographic and video studies:

- Make sure that you find a suitable camera position. Position the camera so that it is above the level of the operation or process that you want to record. Otherwise you may get significant obstructions in the foreground and/or lose information from workers who cannot be seen in the picture frame. Some people recommend at least 15 feet above the operation's level for optimal observations. You can mount cameras on buildings, rooftops, slopes, tower cranes, or specially erected small scaffolds.

- Make sure that you take advantage of the camera's zooming capabilities to capture the entire area of interest.
- The camera needs to be placed in a waterproof box with a window designed to minimize reflections.
- Sometimes it may be difficult to identify the workers whose work you want to follow in a congested area where several crews are collocated. It may be a good idea to ask the workers to wear special color hats or any other distinctive mark for easy identification.
- To frame the recording in its proper context, some people recommend recording a panoramic view of the entire project first and later zoom in to the area of interest.
- When using the editing software, be sure to add information such as the project's name, project's location, operation's name, date and time of the recording, special conditions, and so on to the video.
- The camera should be capable of imprinting each frame with a number and the time.

The detailed analysis of photographic and video studies still requires time, and this is one of the main reasons some managers overlook this in favor of a more informal review of the recordings. However, in order to obtain the maximum benefits possible, a detailed analysis is indispensable. You should also look at the recording at least three to five times, as it is difficult to identify all the interferences, wasted motions, lost time, duplicated efforts, and a variety of other inefficiencies from looking at a recording only a couple of times. You should also study the activities of each crew member or piece of equipment independently.

Activity Sampling

There is no doubt that the best way to evaluate the effectiveness of a worker, a crew, or an entire construction operation or process would be to capture continuous data about the work being carried out and duplicate the operation on a computer simulation. However, such an undertaking would probably be more expensive than the construction operation or process itself. Some day information technologies may allow us to accomplish this inexpensively, but as of today we do not have such capabilities. This should not discourage us, however, as we can take advantage of statistical principles to work with samples and make inferences about the population.

The statistical and probabilistic principles required to scientifically perform activity sampling are out of the scope of this chapter. In other words, we are not going to calculate statistical sample sizes or determine if differences in performance are statistically significant. However, this should not discourage us from taking advantage of recommendations (rules of thumb) based on studies performed in the past where rigorous statistical principles have been applied. This approach is valid when the main objective is to identify potential problems. Once again, we must think of construction as a system of interrelated components. Activity sampling can help us in identifying symptoms of potential company-wide problems. In essence, when performing sampling studies of construction operations you should:

- Have the capability to easily identify those who belong to your sampling group and those who do not. This is especially relevant if the construction operation or process is performed in an area where members working on different operations or processes tend to be collocated. Requiring workers to wear special shirts or hard hats may be extremely helpful in such a situation.
- Perform no less than 400 observations in total. This could mean 40 observations per worker for a crew of 10 workers or 4 observations per worker for a group of 100 workers. This requirement guarantees that a reasonable level of confidence is achieved regarding population representation.
- Have the same likelihood of observing every worker. If you cannot observe some workers because of obstructions or topographic limitations, then your sample may not be representative of the population and the conclusions of your study may not be valid.
- Avoid observations with a sequential relationship. To accomplish this you should collect the observations at random times and in different sequences. A portable computer or a personal digital assistant (PDA) may be used to generate random times and gather observations.
- Annotate the first impression you get when observing a worker. Be careful not to pay attention to what the worker was doing immediately before or what the worker will do immediately after of the observation. The objective of first impressions is to avoid biases.
- Make sure that the basic characteristics of the work situation are the same throughout the observation period. For example, you cannot mix data captured on a sunny day with those obtained on a rainy day.
- Avoid identification by those performing the work. It is a well-known fact that people tend to alter their behavior when they know that they are being observed. Have you ever altered your behavior because you knew you were being filmed? We all have; it is human nature.

> Therefore, you may want to be as inconspicuous as possible. Perhaps you can perform other duties on the site that require your presence there from time to time at unpredictable times. This strategy will make it more difficult for workers to realize that you are observing them and collecting data.

There are three major methods for activity sampling: field ratings, productivity ratings, and five-minute ratings. The particulars of each approach are explained in the following sections.

Field Ratings: Field ratings classify workers only as "working" or "nonworking" at the time of the observations. Working is defined as any engagement in a useful activity. Following are some issues to consider when performing field ratings:

- The use of mechanical counters to record observations may be quite useful. One counter can be used to record people "working" and another one can be used to record all observations.
- Counting requires full undivided attention. You must concentrate while performing counts and avoid any distractions.
- It is not advisable to perform field ratings on the first or the last half hour of a workday or closely before or after lunchtime, unless you want to evaluate how an operation behaves at these particular times.
- In order for an employee to be counted as "working," he or she must be carrying materials, holding or supporting materials, or participating in active physical work (measuring, reading blueprints, writing orders, giving instructions, receiving instructions, operating a machine, etc.).
- An employee should be counted as "nonworking," when he or she is waiting for another task to finish, talking without being actively working, walking around empty-handed, and attending self-operating machines, among other examples.

Productivity Ratings: Productivity ratings go beyond field ratings in the definition of the activities that the workers are performing at the time of the observations. Three main categories are used for classifying work: "effective," "contributory," and "not useful." Therefore, in order to properly apply productivity ratings, one must first define which activities classify into each one of those categories for the construction operation under study. Different construction operations will have different lists of "effective," "contributory," and "not useful" work activities.

In general, "effective" work should include activities directly involved in the operation and are usually carried out in the immediate vicinity where the work is being done. Some examples include installing conduit, pulling cable, energizing a piece of equipment, unpacking materials and equipment, and testing equipment. "Contributory" work includes activities that, while not directly adding to the process of putting together a facility, are still essential for the operation. Examples include handling materials, measuring, reading drawings, building a scaffold, cleanup, giving or receiving instructions, and necessary movements around the work area. "Not useful" work includes all other activities. Some examples include walking empty-handed, taking a coffee break, waiting for materials, waiting for equipment, or waiting for instructions. The difference between contributory work and not useful work may require close attention in some instances. For example, a worker who is waiting for another crew member to pass a piece of material is in a "contributory" stage. However, if there are two workers waiting, then only one of them could be considered in a "contributory" stage. Rework is generally classified as "not-useful" work.

Productivity studies have found that most construction workers spend from 30% to 45% of the time doing "effective" work and another 25% to 35% doing "contributory" work. Therefore, "not-useful" work is experienced between 20% and 45% of the time. This implies that the productivity of a typical construction operation that is not running at optimal levels can probably be improved by at least 25% if proper procedures and techniques are applied. Keep in mind that in addition to reducing "not-useful" work, proper planning and technologies can also reduce "contributory" work to a certain extent. Productivity ratings are very useful as benchmarks to determine the improved performance of a construction operation when different technologies, construction methods, or planning techniques are applied.

As with field ratings, productivity ratings require your undivided attention and should not be performed on the first or last half hour of a work day or close to a lunch break.

Five-Minute Ratings: Five-minute ratings are not considered as statistically valid as field or productivity ratings. This technique is seen more as a shortcut to determine if further evaluation of a particular construction operation is required. Its name is supposed to remind us that we should not observe a crew for less than five minutes. As a rule of thumb, some people recommend to add a minute for each worker in the crew under observation. Therefore, a crew of 6 workers should be observed by at least 11 minutes (5 minutes + 1 minute × 6). The observation period is then broken down into arbitrary but equal subperiods. For example, our 11-minute observation could be broken down into 1-minute sub-periods. Workers are observed in each sub-period and they are classified as "working" or "non-working" as in field studies. However, the key difference here is that the observations are not instantaneous as it is the case for field and productivity ratings. In fact, workers are observed for the full duration of the sub-period. Therefore, they are classified as "working" only if they were working more than 50% of the sub-period according to the judgment of the observer.

In order to get a proper knowledge of a crew's effectiveness, some people recommend performing at least four separate five-minute rating studies in a day.

Notice that productivity ratings do not strictly measure productivity. They do, however, determine the ratio between productive and unproductive time. This is a very valuable piece of information that can be used to evaluate the effectiveness of productivity improvement efforts.

Task Productivity

Task productivity is probably the easiest productivity measure to obtain in the field, but the least useful as well. When determining the productivity of a task, we are really measuring the abilities and proficiency of the worker. Determining how many linear feet of cable can a worker install per hour is useful only as the upper limit of achievable productivity. Projects suffer low productivity not because workers are not producing or working "hard enough." Projects encounter low productivity because of the following:

- *Poor planning of construction operations.* Incomplete or incorrect instructions given to workers, overmanning, stacking of trades, inefficient sequencing, lack of coordination among trades, and shortages of materials, tools, and equipment, among other problems.
- *Design issues.* Inconsistencies among design documents, incomplete designs, and erroneous designs among other issues.
- *Lack of motivation.* Worker suggestions are dismissed, superior performance is not adequately rewarded, and turnover is high, among other issues.

Nonetheless, when new construction tools or technologies are being implemented, measuring task productivity may be very useful as a way to compare the achievable productivity offered by the new methods with that of the traditional approach.

CONCLUSIONS

Improving labor productivity is recognized as an important goal by most contractors. However, some are reluctant to try new construction methods, techniques, managerial approaches, or incentive systems because they do not know how to validate their effectiveness. This chapter discusses practical and simple ways of measuring labor productivity at different levels, ranging from the company to the individual task. Measuring labor productivity before and after applying a new approach can provide valuable information to validate its effectiveness. Furthermore, the ability to measure labor productivity over time allows contractors to establish a continuous-improvement program within their organizations with the aim of enhancing their productivity and profitability.

Contractors may not want to measure labor productivity at all levels. Indeed, it may be more effective if a contractor focuses on properly measuring labor productivity at a few levels, rather than trying to do it at all levels with low reliability and accuracy. Each contractor should evaluate his or her particular circumstances and decide which of the tools presented in this chapter are best suited for his or her company. Furthermore, contractors may also decide to follow a sequential approach, where they measure labor productivity at one level first, before moving to the next one. No matter which approach is adopted, just thinking about productivity improvement and collecting data to evaluate performance is a step in the right direction. The process itself will likely uncover inefficiencies in companywide procedures that, when fixed, may yield significant rewards.

WEB-ADDED VALUE

Measuring.xls

This spreadsheet includes the data and calculations used to create Tables 1.2 to 1.5 and Figures 1.3 to 1.8. Since all formulas are included, you can use this spreadsheet either as a template to enter your own data or as a reference when creating your own spreadsheet.

This book has free material available for download from the Web Added Value™ resource center at *www.jrosspub.com*

2

BENCHMARKING FIELD OPERATIONS

Dr. Thomas E. Glavinich, *University of Kansas*

INTRODUCTION

Benchmarking is a key element in a contracting firm's continuous-improvement program and can be the cornerstone of its quality assurance (QA) program. The goal of field benchmarking is to improve construction quality, productivity, and safety. Effective field benchmarking develops mutually agreed-to performance goals with the crew, designs work processes to achieve these goals, and then sets performance criteria for measuring the effectiveness of work processes. Benchmarking provides the yardstick against which improvement efforts can be measured. Without a method for measuring progress, the contracting firm does not know if improvement efforts are successful. An effective benchmarking program will lead to market leadership and sustained competitive advantage for the firm.

Field benchmarking is not just about setting performance goals and measuring performance toward their attainment. It is about measuring the ability of the contracting firm and its workers to design effective and efficient construction means and methods to get the work completed on time, within budget, and in accordance with the contract documents. The field benchmarking process presented in this chapter requires active crew involvement in the continuous-improvement process, open lines of communication between the contractor and the crew, and, most importantly, trust.

This chapter defines field benchmarking and explains how it can be used to convert crews into self-managing teams, presents the five key elements necessary for an effective field benchmarking program, discusses an eight-step field benchmarking process, presents the "how to" of benchmarking, provides a generic field benchmarking procedure, and discusses organizational learning as an important part of field benchmarking.

FUNDAMENTAL CONCEPTS OF BENCHMARKING

Benchmarking is the continuous and systematic process of identifying best business and construction practices in order to establish performance goals for continuous improvement. Benchmarking can be used to improve any business or construction process engaged in by the contracting firm. Benchmarking is an invaluable tool for effective work process improvement.

The focus of benchmarking is on the business or construction process being improved, not on the outcome. The objective of benchmarking is the identification of best practices and the adaptation of those practices to improve the performance of the firm. The benchmark itself is just the metric used to determine the effectiveness of process improvement efforts. The outcome is the result of the process, and the only way you can improve the outcome is to improve the underlying process.

Many managers mistakenly believe that benchmarking is about identifying the "best number" or benchmark to gauge the effectiveness of a particular business or construction process. Armed with this number, they then set up an internal reporting system that provides periodic feedback from the actual process to compare against the "benchmark." Little, if any, thought or effort is put into improving the underlying process, and when improvements fail to materialize, these managers are quick to condemn and abandon benchmarking. In a way, these managers fit the popular definition of insanity, which is "doing the same thing over and over again and expecting a different result."

Attributes of Effective Benchmarking

An effective benchmarking program must be an ongoing effort that involves everyone in the firm. Benchmarking cannot be just a one-time or one-project effort if the contracting firm expects to obtain long-term benefits from its application. Benchmarking must become an integral part of the firm's day-to-day operations and corporate culture both in the office and at the project site.

Similarly, to be effective, benchmarking must be systematic in its application. There must be a consistent method for analyzing existing business and construction processes, identifying and adapting best practices to improve these processes, measuring the effectiveness of improvement efforts, and finally disseminating and applying what has been learned throughout the firm. An unplanned and poorly executed benchmarking program will produce disastrous results for the firm just like unplanned and poorly executed field operations.

The contracting firm's benchmarking process should be broken down into a set of discrete steps that are accomplished in order. Without a structured approach, the benchmarking process loses its effectiveness. Benchmarking involves management by fact and demands reliable, accurate, and comparable data that can only be gathered and analyzed using a systematic process that everyone understands and adheres to.

Lastly, performance goals or benchmarks must be both realistic and achievable. Benchmarking seeks to identify best business and construction practices. It goes without saying that these best practices cannot be adopted directly but must be adapted to the unique culture and circumstances of the contracting firm. Similarly, the performance goals associated with the adapted process must also be set based on the firm's unique culture and circumstances. If performance goals are not realistic and achievable, the benchmarking program will lose its credibility within the contracting firm's organization and fail.

Benchmarking Strategies

There are four benchmarking strategies that can be used by the contracting firm to improve performance. These four strategies are as follows:

- World-class benchmarking
- Industry benchmarking
- Peer benchmarking
- Internal benchmarking

Each of these benchmarking strategies has the ability to improve the firm's operations. Effective use of these strategies will lead to market leadership and sustained competitive advantage. This chapter focuses on internal benchmarking, but the following paragraphs provide a short description of the other benchmarking strategies that a contracting firm can use to improve its business and construction processes.

World-Class Benchmarking

World-class benchmarking is often referred to as generic benchmarking. In world-class benchmarking, the contracting firm identifies and adapts best business and practices from outside the construction industry. This strategy typically applies to the contracting firm's business processes although there may be generic production practices that could be adapted from other industries to improve construction processes. An example of a business practice that is important to the success of any construction project and lends itself very nicely to world-class benchmarking is the procurement of materials and equipment for installation at the project site.

Industry Benchmarking

Industry benchmarking is sometimes referred to as functional benchmarking. With industry benchmarking, the contracting firm searches for best business and construction practices outside its area of specialization (concrete, sheet metal, electrical, mechanical, plumbing, etc.), but within the construction industry as a whole. Using an industry benchmarking strategy, the contracting firm looks to general contractors and specialty contractors to identify business and construction processes that can be adapted to their way of specialization to improve firm performance. For example, if an electrical contracting firm wants to offer design-build services to its customers, the electrical contracting firm might seek out general contractors who specialize in design-build projects to learn how they market, contract, and perform the work. Successful design-build business and construction practices could be adapted by the electrical contracting firm which will shorten the learning curve, improve the chances of success, and avoid common pitfalls.

Peer Benchmarking

Peer benchmarking is often referred to as "competitive benchmarking"; however, the term is restrictive and not indicative of what this benchmarking strategy is all about. Peer benchmarking seeks to identify and adapt best business and construction practices from within the contracting firm's area of specialization. The term "competitive" leads people to believe that this benchmarking strategy is only concerned with the contracting firm's direct competitors. This is not the case and to take such a restrictive view of this strategy severely limits the benefits that can accrue from its application. In fact, the most effective competitive benchmarking studies are conducted with other peer contracting firms outside of the contracting firm's technical niche or geographic market.

Peer groups are an example of peer benchmarking. Peer groups are composed of noncompeting contracting firms that exchange information and data for continuous improvement. Peer benchmarking is most effectively applied to business and construction practices that are unique to the firm's area of specialization. An example of a peer benchmarking study is determining how electrical contracting firms are successfully diversifying into the growing integrated building systems (IBS) market.

Internal Benchmarking

The fourth strategy is internal benchmarking, which is what this chapter is all about. Specifically, this chapter focuses on field benchmarking, or the internal benchmarking of construction processes. Internal benchmarking is applicable not only to construction process improvement but also to the contracting firm's internal day-to-day business processes such as estimating and bidding.

Internal benchmarking compares business and construction processes within the firm itself. Many day-to-day work processes within a contracting firm are not standardized and differ from person to person and project to project with varying levels of quality, efficiency, and effectiveness. For example, estimators within the same firm may use completely different methods and data to generate bid estimates for similar projects. Internal benchmarking seeks to identify the best practices within the firm and then standardizes these best practices throughout the firm.

Internal benchmarking is very important to the firm's continuous-improvement program. Internal benchmarking results in a thorough understanding and standardization of the firm's internal work processes. Standardized internal processes can then be documented and serve as a basis for both continuous improvement and quality assurance. In addition, effective external benchmarking can only begin after internal business and construction processes are standardized. Peer, industry, and world-class benchmarking can only begin after the internal process being benchmarked is thoroughly understood and documented.

Internal benchmarking also promotes learning within an organization. Just going through the process of internal benchmarking will result in increased communication across organizational boundaries. Heightened organizational self-awareness and enhanced intrafirm communication are both by-products of internal benchmarking.

Importance of Upper Management Support

Field operations are the key to a contracting firm's success. Putting work in place at the construction site is what defines a contracting firm and where it makes its money. Properly applied, field benchmarking will improve construction process quality, productivity, and safety. By their very nature, construction processes are a natural for internal benchmarking.

To be successful and become an integral part of a contracting firm's day-to-day operations, field benchmarking must have upper management support. Everyone within the firm must know that upper management is committed to continuous process improvement in the field and field benchmarking as a means for achieving improved quality, productivity, and safety. This means that upper management must visibly promote field benchmarking, provide project managers with the resources they need to effectively implement field benchmarking, and recognize those project managers who implement field benchmarking on their projects. In addition, upper management must encourage a sharing of benchmarking procedures and information throughout the firm to promote organizational learning, which will not only improve field operations but also estimating, procurement, and other business processes that support the field.

Upper management must also take the lead in working with workers to ensure that everyone understands the purpose of field benchmarking and how it will be implemented in the field. Without the support of the crew, field benchmarking cannot be effectively implemented. Care must be taken to ensure that the proposed benchmarking program and procedures comply with labor agreement terms and conditions and any guidelines for employee participation programs, if these exist. To be effective, upper management must view the construction worker as a partner in its process improvement efforts.

Project Manager as the Catalyst

The project manager is the catalyst that will make benchmarking work in the field. The project manager is the only person in the organization who can effectively convert benchmarking from management concept to useful management tool at the project site. If the project manager views field benchmarking as just another management buzzword or program of the month, it is doomed. The project manager must openly support field benchmarking in both words and actions in order to actively involve the foreman and crew.

The success of the benchmarking program depends largely on the project manager's leadership abilities and commitment to continuous improvement. It is upper management's responsibility to ensure that the project manager understands the purpose and benefits of field benchmarking as well as how benchmarking can be successfully implemented on his or her project. This requires initial and ongoing training as well as the unwavering support and encouragement by upper management throughout the implementation process. Finally, upper management must be ready and willing to recognize and reward the project manager's continuous-improvement efforts.

Foreman's Role in Benchmarking

The foreman faces the greatest challenge when implementing field benchmarking. If field benchmarking is to be successful, the foreman's role must change from one of "boss" to one of facilitator and coach. This shift in role is necessary because an effective field benchmarking program requires that construction workers take responsibility for their own work and make many of the routine decisions about how the work will be performed. In this new role, the foreman spends most of his or her time planning and scheduling the work, which includes making sure that the workers have the overall guidance, materials, and tools available when needed. In addition, the foreman also gathers production data and provides feedback to the project manager and the workers so that everyone knows how the project is progressing and has a basis for continuous improvement.

Crews Become Self-Managing Teams

A crew is a working group where the focus is on individual performance. Without benchmarking, the focus on collective crew performance is sometimes lost in the field. The foreman directs the work, and individual crew members carry out the foreman's instructions to the best of his or her ability. Each crew member is evaluated by the foreman based on his or her performance and not the collective effort of the crew.

Construction is a team effort. Very few activities on a construction site are solo activities, and even then, these solo activities typically depend on other work that has already been done. Everybody's work depends on the work of others, and this is especially true at the crew level. Success in construction is a product of team work and not individual effort. Transforming a crew of individual workers into a team is very important because a crew is no better than the individual bests of its members. A team, on the other hand, has synergy and the team best is always greater than the sum of individual crew member bests.

A crew is a natural team in the rough. Teams are made up of individuals with complementary skills that can be focused on achieving a common objective. Crews can easily evolve into teams through field benchmarking because it involves everyone in the planning, performance, and evaluation of work processes. The benchmarking process requires that a crew establish a performance objective for an activity, design a construction process that will achieve the performance objective, and then evaluate the effectiveness of the construction process based on the crew's performance as a whole.

KEY FIELD BENCHMARKING ELEMENTS

To have a successful field benchmarking program, five key elements must be in place:

- Crew trust
- Crew involvement
- Crew empowerment
- Crew stability
- Crew training

Each of these key elements is important to the success of the field benchmarking program and dependent on the other four elements. Lacking any one of these elements will seriously jeopardize the chances of success and effectiveness of the field benchmarking program. This section discusses each of these five key elements and what a contracting firm needs to do to ensure that each of these key elements is in place.

Developing Crew Trust

For field benchmarking to be successful, first and foremost there must be trust between the crew and the contracting firm. The construction worker must be confident that the purpose of field benchmarking is continuous improvement of construction processes and not the evaluation of individual or crew performance. Actions speak louder than words, and the project manager and foreman are key to building trust in the field. If the contracting firm's challenges are to be the worker's challenges, then the worker's problems and concerns must become the contracting firm's problems and concerns. Without a mutual understanding and concern for each other's challenges and problems, trust cannot exist.

The sharing of information is an important part of building trust. For field benchmarking to work, the contracting firm must share planned worker-hour information with the crew so that they can make an informed decision about what the benchmark for a particular activity should be. Similarly, the contracting firm also needs to share planned construction methods, production rates, and activity schedules with the crew so that they understand what needs to be done and why. The crew cannot effectively participate in construction process improvement without all of the pertinent information.

On the other side of the coin, the construction workers must trust the contracting firm and provide accurate production data on a daily basis. "Banking" production on good days to cover bad days undermines the benchmarking process and makes it impossible to get the feedback needed to ensure that the process being benchmarked is on track. The worker must firmly believe that the daily production data turned in will be used for process evaluation and not for judging individual or crew performance.

Similarly, once the production data is turned in by the workers, the contracting firm needs to compile the data and provide the results to the workers accurately and honestly. In order to be an effective partner in construction process improvement, the worker needs to know what is working and what is not working as soon as possible. Without timely feedback, the worker will feel alienated from the process and will be unable to contribute his or her expertise to process improvement. The worker has the best understanding of how a construction process actually works and how it can be improved.

Encouraging Crew Involvement

To build an effective team, everyone must be involved in the day-to-day planning and execution of the work at the project site. Each member of the crew has unique capabilities, experience, and perspective to contribute to the team's continuous-improvement efforts. Many times the newest and most inexperienced team members have the best ideas because they are not constrained by habit or tradition.

In order to get the best from the team, everyone must be encouraged to take an active role in the improvement of construction processes and the solving of specific problems that arise. Taking an active role in planning and problem solving promotes ownership of the construction process and an understanding of why an activity is performed as it is. This leads to greater support from the crew and an increased dedication to the successful completion of the project.

One of the best ways to get everyone involved is through regularly scheduled crew brainstorming sessions. These sessions should focus on planning an upcoming activity or solving a particular problem. Brainstorming sessions should include everyone whether they are involved directly in the activity or not. During these sessions everyone's ideas are taken seriously and considered on an equal basis. The best course of action to come out of a successful brainstorming session is usually not any one person's idea but a combination of ideas from several participants. Brainstorming promotes teamwork, and the time invested pays dividends through higher morale and greater productivity.

The crew should be involved in the project planning process as soon as possible. The greatest opportunity for construction process improvement occurs during the early stages of the project, which is when the experience and expertise of the crew can have the biggest impact on the project outcome. During the project planning process, the crew should become familiar with the plans and specifications and be consulted on means and methods as well as on planned materials and equipment and project staging. Not only will this provide the worker a basis for assisting in construction process improvement, but it will also provide him or her with an appreciation of the constraints within which the contracting firm is working.

The development of a project mission statement early in the project can also promote greater crew involvement. A project mission statement should address the project directly and challenge the crew in some way. Unlike the typical corporate mission statement that deals with a going concern and is qualitative, the project mission statement must recognize the finite nature of the project and be measurable to be effective. The crew must be instrumental in developing the mission statement, and the statement must be meaningful to the crew. Usually, the best project mission statements result from the crew's direct involvement in the project planning process. An effective project mission statement might be something as simple as "zero punchlist."

Facilitating Crew Empowerment

As discussed in the previous section, field benchmarking is the catalyst that transforms crews into self-managing teams. Self-managing teams demand worker empowerment. To get the best from the crew, the foreman must be willing to allow the crew to plan and execute the work in their own way. This requires trust and the belief that individual crew members want and can do the job at hand. Empowerment allows the worker to bring his or her experience, knowledge, and creativity to bear on construction process improvement. The foreman must work with the crew to set specific performance goals, ensure that the crew has what it needs to complete the work, and then get out of the way and let the crew perform the work. Crew empowerment will lead to higher morale, increased job satisfaction, and improved productivity.

Empowerment means trusting workers to do their own work and pushing decision making down to the lowest level. Rather than the foreman giving orders, workers should be assigned tasks and areas. It is then up to the worker to decide how best to get the job done within the parameters established by the foreman.

For the project manager and foreman, this is probably the most difficult of the five key elements. It requires letting go and giving up control of many day-to-day job details. It is a change from the way jobs have been run historically and challenges traditional paradigms. However, like the other four key elements, without empowerment, field benchmarking will not work.

Committing to Crew Stability

Crew stability is a key element in transforming crews into self-managing teams. It is important to maintain a steady crew throughout the project in order to build an effective and cohesive team. Constant changes in crew personnel over the course of a project will stymie the team-building process. Crew stability requires commitments from the foreman, workers, and contracting firm. The foreman must spend more time planning and coordinating work in order to level the manpower demands over the life of the project. The workers must understand that there will be times when the workload is greater than what the size of the crew would normally dictate. Similarly, the contracting firm must also realize that there will be times when the workload is less than the size of the crew would otherwise dictate. Commitment to crew stability by the foreman, crew, and contracting firm can only occur if all three believe that benefits will accrue from the team approach.

On almost any project it will be impossible to have a perfectly stable crew from start to finish. In the beginning of the project, new workers will be hired as the work increases. Similarly, as the project comes to a close and the workload decreases, workers will need to be laid off. It is very important that everyone knows from the beginning the procedure for determining who is laid off and in what order as the project winds down. This procedure must be perceived as being fair, and if field benchmarking is to work, under no circumstances should the reported daily individual or crew productivity be used as the criteria for determining layoffs.

Providing Crew Training

In order to effectively participate in a team environment and contribute to the continuous-improvement process, the workers must have an understanding of quality and continuous improvement, how teams are formed and operate, and how to analyze and improve construction processes. This may require training that can be performed at the jobsite by a qualified member of the contracting firm's organization or an outside facilitator. Crew training will require an investment by the contracting firm and a commitment to the team approach by the foreman and crew. Again, for the team approach to be successful, everyone must believe that significant benefits will accrue from the investment of time and money.

FIELD BENCHMARKING PROCESS

This section presents a detailed step-by-step process for field benchmarking. It is important for the contracting firm to have a structured process in place that everyone understands and adheres to so that data gathered from different crews and different projects are comparable. Without a structured field benchmarking process, a lot of the potential for process improvement and overall organizational learning will be lost and the ability of field benchmarking to improve the contracting firm's competitive advantage will be severely limited.

The field benchmarking process is illustrated in the flowchart shown in Figure 2.1 and includes the following eight steps:

Step 1. Define the work
Step 2. Set the benchmark
Step 3. Preplan the work
Step 4. Implement the preplan
Step 5. Gather production data
Step 6. Chart progress
Step 7. Evaluate and audit progress
Step 8. Continue implementation

The following paragraphs describe each of these steps.

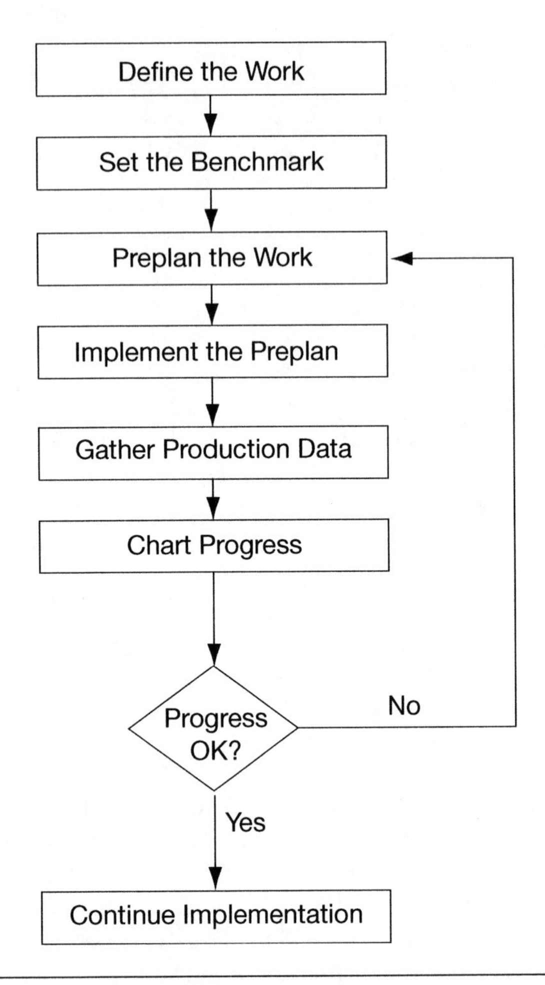

Figure 2.1 Field benchmarking process.

Step 1: Define the Work

The first step in the benchmarking process is to define the activity to be benchmarked. Any activity can be benchmarked; however, the best candidates are those that are very labor-intensive. Activities that are especially good for field benchmarking are very repetitive, are judged to be either complex or difficult, or are on the schedule critical path. Because these activities have the greatest potential for improvement, they will benefit the most from benchmarking. Remember, benchmarking requires an investment of both time and effort of everyone involved. Therefore, the required investment should only be made in those activities that have payback potential. Selecting and defining activities to be benchmarked is part of the art of field benchmarking.

When defining an activity to be benchmarked, keep in mind that all the work needed to complete the activity needs to be included. Also, when defining the work, remember that the crew will be keeping track of its daily effort. Make sure that the activity is well defined and easy for the crew to track. This is especially true when there are multiple crews performing the same work. Consistency in data gathering is critical to successful field benchmarking.

An example of an activity to be benchmarked might be the installation of 3,000 linear feet of 6-inch cable tray. If, the crew installing the cable tray will be installing the cable tray hangers at the same time as the cable tray and associated fittings, then the scope of the activity being benchmarked should include the installation of cable tray, fittings, hangers, material handling, and layout. If on the other hand, the hangers are to be installed first and then later in the project the cable tray will be installed to avoid conflicts with other trades, then the installation of the hangers should be benchmarked separately.

Step 2: Set the Benchmark

Once the activity to be benchmarked has been defined, it now needs to be analyzed in light of existing conditions at the site. The actual conditions at the site may vary considerably from what was anticipated at the time that the bid estimate was put together. Assumptions about job conditions made during the estimating process may or may not be valid, and productivity may be higher or lower than anticipated.

The analysis of the work activity should be done with the crew that will be performing the work. It is important for the crew to set the benchmark so that it will be their goal. Benchmarks should be set with the crew considering the estimated worker hours, perceived difficulty of the work, and other factors. In addition, a benchmark must be perceived as being realistic and achievable or no one will take it seriously.

Typically, the benchmark should be set in terms of worker hours per unit being installed. In the case of the 6-inch cable tray discussed above, the benchmark could be expressed in terms of worker hours per hundred feet of installed tray, which includes the installation of the tray, fittings, hangers, material handling, and layout. If the hangers are to be installed separately, then they could be benchmarked in terms of worker hours each.

As noted, the benchmark should be expressed in terms of worker hours per unit installed. The benchmark should not be expressed in terms of cost or total quantity installed. This is because the project manager, foreman, and crew have no control over the estimator's pricing or the market fluctuations that affect the final buyout. Further, the project manager, foreman, and crew may have no control over the final total quantity as a result of field routing to avoid interferences and other factors. It is critical that the activities being benchmarked are under the complete control of the workers responsible for performing the work.

In the 6-inch cable tray example, the estimated unit productivity for its installation including hangers on this project might have been 18.0 worker hours per hundred linear feet installed. However, knowing the project and formulating the installation plan around its capabilities, the crew may decide that the unit productivity for tray installation should be 16.5 worker hours per hundred linear feet installed. The 16.5 worker hours per hundred linear feet installed would be the crew's benchmark for this activity.

The natural tendency is to use the bid estimate worker hours as the upper bound on the activity benchmark. This is understandable because the bid estimate worker hours establishes the initial budget for the project against which overall project profitability is measured. However, the bid estimate worker hours should not constrain the setting of the benchmark if the field benchmarking process is to be credible. Granted, the bid estimate worker hours are very important, but as noted above, actual existing conditions at the site must be factored in for better or worse.

Step 3: Preplan the Work

The next step in the field benchmarking process is to preplan the work. In preplanning the work, the crew determines the construction means and methods along with the labor, materials and installed equipment, and tools and construction equipment needed to meet or exceed the benchmark set in Step 2. Again, it is very important that the crew be involved in determining how the work will be performed because this promotes ownership of the construction process and takes advantage of the workers' experience and creativity. The goal is to work smarter, not harder.

An example activity preplanning form is provided in Figure 2.2. A good preplan should answer the following questions about the activity being benchmarked:

- What needs to be done?
- How should it be done?
- When should it be done?
- Who should do it?
- What is needed to do it?

The preplanning process should result in a detailed plan for how the work will be accomplished. This is where brainstorming becomes important. The crew should plan the work as a group and should not be constrained by the estimate

PROJECT:			
PROJECT NO.		PLAN BY	
ACTIVITY DESCRIPTION			
FOREMAN:		PLAN DATE	
SCHEDULED START	DATE:	EXPECTED FINISH	DATE:
	TIME:		TIME:
CREW:			
DRAWING/SPECIFICATION REFERENCES:			
MATERIALS/EQUIPMENT/EXPENDABLES REQUIRED			QUANTITY
TOOLS AND EQUIPMENT NEEDED			QUANTITY
WHEN FINISHED WITH THIS ACTIVITY:			

Figure 2.2 Activity preplanning form.

with regard to either materials or equipment used. Preplanning should address not only material installation but also material handling and other activities that can impact productivity.

Step 4: Implement the Preplan

Once the preplan has been developed, the next step is for the crew to implement the preplan in the field. In this step the crew organizes itself and carries out its work in accordance with the preplan. This is where the foreman takes on the roles of facilitator and coach rather than the traditional role of "boss."

Step 5: Gather Production Data

As the work progresses, production data needs to be gathered to determine if the preplan is working. Everyone gathers and reports his or her own production data, but only the total crew production data is used when analyzing progress toward achieving the benchmark. It is very important that the gathered data be used only to evaluate the effectiveness of the preplan and not the performance of the crew or individual worker. This is the only way that trust can be built, which will lead to the accurate reporting of data and a meaningful dialogue on construction process improvement.

Daily production data should be gathered with regard to what work was performed during the day, problems and delays encountered, and suggested construction process improvements. The daily activity production data from each crew member should be summed to get the total daily production data for the work activity. The production data provided by each crew member should then be destroyed immediately because the focus is on the process and overall process productivity and not an individual crew member productivity.

Step 6: Chart Progress

Progress Tracking

Progress toward meeting and surpassing the performance benchmark should be tracked on a daily basis. Measurement is the foundation of the continuous-improvement process. Without ongoing tracking, there is no way of knowing if the construction process is meeting quality, productivity, and safety expectations. There are a number of ways to track progress, but the simplest and most effective method is the run chart. A run chart (also called a trend chart) provides a timeline for comparing data about a variable of interest. In the case of field benchmarking, the variable of interest is usually productivity or the amount of materials installed per unit time.

Run Chart Data and Calculations

Run charts can easily be generated using spreadsheet software. Figure 2.3 illustrates a spreadsheet that could be used for gathering production data and for performing the calculations needed to generate a run chart for the cable tray example. As can be seen from Figure 2.3, the number of worker hours spent on the activity being benchmarked and the installed quantity of cable tray are gathered from the crew and recorded daily. From this data, a cumulative total worker hour effort and cumulative installed quantity can be calculated. Daily unit productivity can then be calculated using the daily effort and daily installed quantity. Similarly, the average unit productivity can be calculated on a daily basis using the total effort and total installed quantity.

The following paragraphs describe the benchmarking information and calculations shown in Figure 2.3.

Activity Description:			6-inch Cable Tray Installation			
Activity Scope:			Install 3,000 linear feet of 6-inch ladder cable tray including supports, fittings, materials handling, and layouts			
Estimated Rate:			18.0 MH/CLF			
Activity Benchmark:			16.5 MH/CLF			
	EFFORT		**INSTALLED QUANTITY**		**UNIT PRODUCTIVITY**	
ACTIVITY DAY	**DAILY (MH)**	**TOTAL (MH)**	**DAILY (LF)**	**TOTAL (LF)**	**DAILY (MH/CLF)**	**AVERAGE (MH/CLF)**
1	24	24	118	118	20.3	20.3
2	20	44	108	226	18.5	19.5
3	24	68	129	355	18.6	19.2
4	20	88	97	452	20.6	19.5
5	16	104	79	531	20.3	19.6
6	12	116	62	593	19.4	19.6
7	12	128	64	657	18.8	19.5
8	12	140	69	726	17.4	19.3
9	12	152	73	799	16.4	19.0
10	12	164	73	872	16.4	18.8
11	12	176	62	934	19.4	18.8
12	16	192	120	1,054	13.3	18.2
13	16	208	113	1,167	14.2	17.8
14	16	224	108	1,275	14.8	17.6
15	16	240	115	1,390	13.9	17.3
16	16	256	117	1,507	13.7	17.0
17	8	264	54	1,561	14.8	16.9
18	8	272	53	1,614	15.1	16.9
19	8	280	60	1,674	13.3	16.7
20	8	288	58	1,732	13.8	16.6
21	8	296	48	1,780	16.7	16.6
22	16	312	100	1,880	16.0	16.6
23	24	336	162	2,042	14.8	16.5
24	24	360	180	2,222	13.3	16.2
25	24	384	186	2,408	12.9	15.9
26	16	400	122	2,530	13.1	15.8
27	16	416	111	2,641	14.4	15.8
28	16	432	112	2,753	14.3	15.7
29	24	456	124	2,877	19.4	15.8
30	24	480	123	3,000	19.5	16.0

Figure 2.3 Field benchmarking spreadsheet.

Activity Description: The activity description provides a concise description of the activity being benchmarked. The activity description should be the same description that was used in developing the project estimate and is being used in both the project cost reports and schedule.

Activity Scope: The scope of the activity describes what is included in the activity. In the case of the cable tray example, the activity scope included the installation of 3,000 linear feet of 6-inch ladder cable tray as well as the installation of supports and fittings. Material handling and layout was also included as part of this activity. The activity description is very important so that what is involved in the activity is clearly understood in setting the benchmark, preplanning the work, evaluating progress, and, finally, using the information generated to improve the estimating process on future projects.

Estimated Rate: This is the production rate that was used to generate the original project estimate. This is important because it provides a basis for starting to develop the project benchmark as well as provides important data for continuous improvement of the estimating process. The estimated productivity rate for this example is 18.0 worker hours per 100 linear feet of 6-inch ladder cable tray installed.

Activity Benchmark: The activity benchmark is the production rate that the crew has set for itself based on the conditions on-site when the activity is going to be carried out. Conditions may be better or worse than what was anticipated when the estimate for this activity was developed. Therefore, the activity benchmark may be smaller or greater than the estimated production rate. In this example, the crew believes that it can achieve an average overall productivity rate of 16.5 worker hours per 100 linear feet of cable tray, which is about 8% better than the estimated productivity of 18.0 worker hours per 100 linear feet.

Activity Day: Each day work is performed on this activity it is recorded. In the example, days are numbered from 1 through 30. In actuality, the activity day should be a date that corresponds with activity reported on the daily job report.

Daily Effort: Daily effort is the total number of worker hours spent on this activity on a particular day. As can be seen from the example, 16 worker hours were expended on this activity on Day 5. The nature of the work done that day and who is doing it is not known from the chart and would have to be determined from the daily job report. There is no differentiation between the classification of workers that have charged worker hours to this activity and could include apprentice, journeyman, and working foreman hours. Only worker hours of effort are counted here because crew makeup is often out of the control of those doing the work.

Total Effort: Total effort is just the cumulative number of worker hours invested in this activity to date. For example, the total effort for Day 5 is 104 worker hours, which represents the sum of the daily effort worker hours from Day 1 through Day 5.

Daily Installed Quantity: The daily installed quantity is the total number of linear feet of cable tray installed on a particular day. From the spreadsheet, it can be seen that 79 linear feet were installed on Day 5. No credit is given until the cable tray is complete.

Total Installed Quantity: Total installed quantity represents the total number of linear feet of 6-inch ladder cable tray that has been installed to date. The total installed quantity for Day 5 in the example is 531 linear feet, which represents the sum of the daily linear feet installed from Day 1 to Day 5.

Daily Unit Productivity: Daily unit productivity measures the unit productivity per day and is calculated by dividing the daily installed quantity by the daily effort. The daily unit productivity for Day 5 is calculated by dividing the daily effort of 16 worker hours by the daily installed quantity of 79 linear feet, or 20.3 worker hours per 100 linear feet.

Average Unit Productivity: The average unit productivity measures the running average productivity for this activity. The average unit productivity is calculated for Day 5 by dividing the total effort of 104 worker hours by the total installed quantity of 531 linear feet, which results in 19.6 worker hours per 100 linear feet. The average unit productivity is greater than both the estimated and activity benchmark productivity rates, but the important consideration is not the daily rate but the overall trend, which is downward.

Run Chart Construction

Figure 2.4 provides the run chart for the cable tray installation. As the figure shows, a run chart is simply a line graph that plots unit productivity against days worked on the activity. Run charts are easily understood, and trends in productivity can easily be seen and analyzed. The plot can be daily unit productivity, a running average unit productivity, or both. Both plots are useful and are shown in Figure 2.4. The daily unit productivity plot will highlight both favorable and unfavorable trends more quickly than the running average unit productivity plot. However, the running average unit productivity plot should be the focus because it is the comparison of the overall average unit productivity to the benchmark that is meaningful and not daily productivity swings.

Figure 2.4 shows that daily productivity can be very erratic depending on the actual work performed or conditions encountered by the crew on a particular day. Remember that the activity being benchmarked is the installation of cable tray. Therefore, if the crew spends a significant amount of a day on layout, material handling, or just installing supports with little actual finished tray installation, daily productivity will be low based on a measure of actual tray installed. On the other hand, following days may show a very high productivity because a lot of the

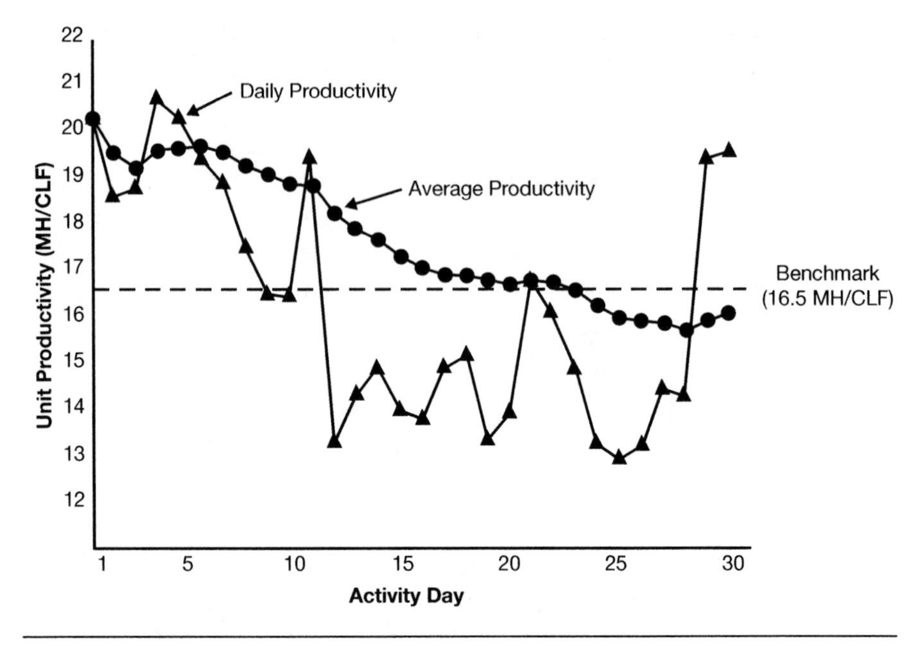

Figure 2.4 Field benchmarking run chart.

preparation work has been done and the crew can focus on installation of the tray itself. This may explain the productivity swing in Figure 2.4 between Day 11 and Day 12. The average activity curve levels out these daily productivity swings and gives a truer indication of how the activity is actually progressing.

Also shown in Figure 2.4 is the benchmark of 16.5 worker hours per 100 linear feet of tray installed. As the tray installation progressed, the crew's average unit productivity improved and the activity completed with an average unit productivity of 16.0 worker hours per 100 linear feet of tray installed. The crew was able to beat the benchmark it had set for itself. Completing the activity in 480 worker hours as opposed to 540 worker hours provides 60 worker hours that may be needed on another activity whose benchmark may be set above the estimated unit productivity due to unanticipated field conditions, an estimating error, or some other reason.

Step 7: Evaluate and Audit Progress

Progress needs to be evaluated on an ongoing basis to identify trends and take corrective action as required. Again, what is being evaluated is the effectiveness of the preplan for achieving the benchmark and not the performance of the crew or individual worker. If there are no problems and progress is being made toward the achievement of the benchmark, then work should continue in accordance with the preplan. If there are problems or the crew notes ways in which the construction process can be improved to achieve even greater quality, productivity, and safety, then the process should be modified accordingly.

Activity benchmarking data must be audited on a regular basis to ensure that it is accurate. Total labor hours expended to date should be compared to the job cost report labor hours to ensure that there is no significant discrepancy. Similarly, the total quantity of materials installed should be compared against the actual material delivered to the job less jobsite inventory to ensure that there is no significant discrepancy. Any significant discrepancies should be corrected and the benchmarking summary spreadsheet and run chart corrected as required.

Step 8: Continue Implementation

As noted, if there are no problems, the crew should continue installation in accordance with its preplan. This is also where the project manager should evaluate the estimated unit productivity and the construction means and methods being used. If it is apparent that the estimated unit productivity was too low, the project manager should share this information with the estimator so that more accurate estimates can be generated in the future. Similarly, if the crew has developed an innovative method for performing the work that could benefit other projects, the project manager should share this information so that the productivity gains realized on this project can be realized on others. In either case, communication of benchmarking information within the contracting firm provides the opportunity for organizational learning and increased competitive advantage.

IMPLEMENTING FIELD BENCHMARKING

This section provides an eight-step field benchmarking process and describes each step of that process. Frequently asked questions (FAQs) about the mechanics of implementing an effective field benchmarking program serve as the basis for this section. The FAQs that are answered in this section are as follows:

- How do you select activities to be benchmarked?
- How do you define the work?
- How do you set the benchmark?
- How do you preplan the work?
- How do you gather production data?
- How do you chart progress?
- How do you evaluate progress?
- How do you communicate progress?
- How do you audit production data?

How Do You Select Activities to be Benchmarked?

Construction projects are made up of a myriad of discrete activities that need to be completed in order to finish the project. Every construction activity has the following five characteristics:

- Activities consume time
- Activities consume resources including labor, materials, and equipment
- Activities have a definable start and finish
- Activities are assignable
- Activities are measurable

As a result, any construction activity within the contracting firm's scope of work can be benchmarked. When the concept of field benchmarking was first applied in the field, it was thought that every activity should be benchmarked. Activities were usually defined by line items in the project estimate, project budget, or cost report, and sometimes by the project schedule. It was quickly found that benchmarking takes time and effort on the part of the entire project team, and it was neither practical nor productive to attempt to benchmark each individual activity. Treating an activity such as "electrical trimout" with the same importance as a key activity like the installation of floor duct that is very labor-intensive and must be carefully coordinated with the slab-pour schedule trivialized the field benchmarking process and reduced it to a data-gathering exercise. Attempting to benchmark each activity on a construction project should be avoided because it will undermine the project team's enthusiasm for field benchmarking and what it can accomplish. Instead of being a team-building initiative that draws on the knowledge and creativity of the project team, field benchmarking becomes just another tracking system that will overwhelm the project manager, foreman, and crew.

All construction activities are not equally important to project success even though every activity needs to be completed in order to finish the project. On every construction project there are a handful of activities that are critical to the success of the project. These few critical activities are the construction activities that should be benchmarked; all other activities should be tracked by traditional methods such as the schedule and periodic manpower and cost reports comparing actual to planned expenditures.

Despite the emphasis on measurement and analysis of data, field benchmarking is an art rather than a science. Part of the art of field benchmarking is the identification of the handful of construction activities that must be completed successfully in order to successfully complete the project. Risk should be the criterion for selecting individual project activities to be benchmarked. Activity risk can result from concern about safety, labor productivity, installation quality, activity budget, activity schedule, and interface with other trades, among other parameters as well as a combination of parameters. The first step in the effective use of field benchmarking in a construction project is identifying those few activities that have the potential to make or break the project and making them the focus of the field benchmarking effort.

Experience has shown the best candidates for benchmarking are those activities that are very repetitive, are judged to be either complex or difficult, or are on the schedule critical path. Usually, the most labor-intensive construction activities should be selected for benchmarking because these activities usually have the greatest potential for improvement and will benefit most from the project team's process improvement efforts. Activities that are equipment-intensive will not benefit as much from field benchmarking because the type of equipment used, the number of units available, and the layout of the work drives productivity, rather than the worker's effort. Construction process simulation and analysis are more effective than field benchmarking for equipment-intensive construction projects.

Based on the criteria and recommendations outlined above, the project manager should review the project contract documents, including the plans and specifications and the project estimate and schedule, to identify those activities that he or she deems critical to the success of the project. The number of activities to be benchmarked should be less than 10, and it is best if the number is around five or six. Focusing on activities that are truly critical to project success will focus the construction team on what is important. Usually, getting these few critical activities right will result in everything else falling into place. Keep in mind when selecting construction activities to be benchmarked that "fewer is always best."

How Do You Define the Work?

Another important part of the art of field benchmarking is defining the work or activity that will be benchmarked. Defining the activity is as important as selecting that activity, and defining the scope can have a significant impact on the success of the benchmarking effort. Narrowly defining an activity may restrict the possible process improvements that are identified and implemented, which in turn may reduce the amount of improvement possible. In other words, restricting the definition of an activity to include only the installation of material at the jobsite could result in important process improvements being missed such as fabrication and bundling of materials by the manufacturer or distributor, material delivery and handling, and supply and use of innovative tools and equipment.

The project manager should define the work as precisely as possible because this definition will set the boundary and determine what processes make up the construction activity and are subject to measurement and improvement. For instance, the project manager needs to determine if the construction activity being benchmarked should include the procurement and material delivery processes or if only the material handling and installation processes at the jobsite are to be considered. Also, the project manager must determine what constitutes the installation. Suppose, for instance, that the installation of busway is to be benchmarked. Does the busway installation include the installation of hangers, or is that another activity? Defining the work to be benchmarked is integral to the benchmarking process and must be performed by the project manager with care.

How Do You Set the Benchmark?

Setting the benchmark for a selected construction activity is an important part of the field benchmarking process and needs to involve the entire construction team. There are six key principles to setting an effective benchmark:

- Involve everyone in the process.
- Build trust by sharing information.
- Use bid information as the starting point.
- Consider historical and outside information.
- Factor in environmental and site conditions.
- Set an achievable benchmark.

Involve Everyone in the Process

The entire project team should be involved in setting the activity benchmark that will be the basis for formulating the activity preplan and measuring success. The benchmark should be set as a group that includes the foreman and the crew and is facilitated by the project manager. Everyone must be involved in setting the benchmark or there will not be any "buy-in" on the part of the foreman and crew, and without ownership, there will be no incentive to achieve the benchmark.

Build Trust by Sharing Information

Successful field benchmarking requires trust, and an important element of building trust is the sharing of information. Information sharing is vital in setting the benchmark, and at a minimum, it should include the number of worker hours estimated for the construction activity being benchmarked, the quantity of work estimated, the resulting estimated average productivity in terms of worker hours per unit work put in place, and any assumptions made in estimating the number of worker hours and quantity of work. To get the best from the foreman and crew, they must feel a part of the process, and setting an achievable benchmark with their input is critical.

Use Bid Information as the Starting Point

Bid information should just be the starting point for establishing the benchmark for a construction activity. It is a mistake to establish the benchmark as the average unit productivity rate by simply dividing the estimated number of worker hours for the activity by the estimated quantity of work. First, the estimate was not prepared by the foreman or crew that will be responsible for performing the work and therefore they have no stake in meeting the estimated unit productivity. Second, the estimated unit productivity was set based on the project contract documents before the project ever started and actual field conditions may be considerably different at the jobsite than what was assumed when the project bid estimate or budget was developed. Third, just using the estimated unit productivity assumes that whoever performed the estimate knew the right productivity rate and ignores the fact that the actual crew working on the project probably has a better feel for what the unit productivity should be, given their capabilities and jobsite conditions.

Consider Historical and Outside Information

In setting an activity benchmark, other information besides the bid estimate or project budget for the activity should be considered, including past experience with similar activities under similar jobsite conditions. Productivity studies can also be used to adjust planned productivity rates for anticipated jobsite conditions. In addition, if the contracting firm and project team do not have experience installing a particular material or equipment, manufacturer information about the installation of a particular material or equipment should also be obtained and factored into the benchmark.

Factor in Environmental and Site Conditions

As noted above, the bid estimate and project budget are put together before the start of the project, and at that time, actual environmental and site conditions are not known. Actual field conditions need to be factored into the benchmark for it to be realistic. The foreman and crew can provide invaluable input as to the conditions they are encountering on a day-to-day basis at the jobsite and how these conditions will impact productivity on the benchmarked activity either positively or negatively. If actual environmental and site conditions are not factored into the setting of the activity benchmark, then the benchmark may be unrealistic or not challenging enough for the crew. Either way, the benchmark under these conditions will be unrealistic and a demotivator. Properly set, the activity benchmark should be a realistic and motivate the crew to do its best work.

Set an Achievable Benchmark

Whatever activity benchmark is set, it must be an achievable benchmark that reflects the physical realities that the crew is encountering on the project site. An unrealistic or unachievable activity benchmark will at best be ignored by the crew and at worst actually be a demotivator. If the estimated unit productivity rate is unrealistic given actual jobsite conditions, the project manager should not attempt to impose this benchmark on the crew. If it was an estimating error in fact or in judgment, the project manager should own up to it and work with the crew to minimize the unfavorable variance that will likely result from the error. On the other hand, if the problem is due to forces outside the control of the crew such as poor job coordination resulting in crowding, disruption, stacking of trades, acceleration, or other impacts, the benchmark should reflect these conditions and the crew should not be responsible for them. Instead, the contracting firm should proceed with its work at the jobsite as best it can and seek reimbursement for delays and lost productivity due to others' actions through the change order process.

How Do You Preplan the Work?

Once the activity benchmark is set, the entire project team should plan how the work will be accomplished. As in setting the benchmark, the project manager should facilitate the development of the preplan with the foreman and crew. The activity preplanning session will typically involve brainstorming to determine the best means and methods to complete the activity given the activity benchmark and actual jobsite conditions. The project team should think outside the box when developing the activity preplan, and other people such as the estimator, procurement and safety personnel, manufacturer and distributor personnel, and engineers should be involved when appropriate.

The outcome of the brainstorming session should be a written activity preplan. Figure 2.2 provides an example preplanning form that could be used for documenting a preplan. There is no right or wrong format for preparing a written preplan for a construction activity. The contracting firm should develop its own form that can be either paper-based, electronic, or a combination that meets its needs based on the type of work that it does. The important consideration is that preplanning is standardized and that information included in the written preplan is consistent so that crews know what to expect and how to read the preplan.

There is a tendency for the project manager to delegate preplanning to the foreman who will be responsible for managing the day-to-day work on the benchmarked activity. This is a mistake because it sends the message to the foreman and crew that the project manager does not think benchmarking is important and that it is not expected to have a significant impact on the project outcome. Actions speak louder than words, and the project manager needs to be an active participant in developing the activity preplan. In addition, the activity preplan may need the project manager's assistance in getting a material specification or code variances, assigning more workers or workers with different skills, obtaining more or different materials and equipment, among other items. For example, the project specification may require drilled anchors where shot anchors will work equally well. In order to make this change, the project manager will need to put together a request and get permission from the design team before proceeding.

How Do You Gather Production Data?

For field benchmarking to work, trust is essential. Each worker must track his or her own productivity and report it on a daily basis. Workers must believe that the production data they gather throughout the day will be used solely for construction process improvement and not for personnel evaluation and layoff decisions.

The contracting firm should standardize the method for gathering production data so that foremen and workers will know what to expect from project to project. It is best if the contracting firm provides the means for the worker to record the work performed on a particular day as well as the amount of time spent and the actual units put in place. This can be accomplished by developing a form to record this information on a 3-by-5-inch card that can be easily carried and filled in by the worker. A simple pocket-sized notebook that workers can carry to write down the needed benchmarking data along with obstacles encountered and ideas for construction process improvement also works.

How Do You Chart Progress?

Progress should be charted using a spreadsheet program with graphing capabilities. Whenever possible, the data should be entered and processed at the jobsite by the foreman to provide near-instantaneous feedback on how benchmarked activities are progressing. The daily progress should be sent electronically to the project manager and anyone else who needs the information but is not on-site. If the foreman is not able to enter and process the benchmarking data daily, then the data should be sent back to the office where it can be entered and processed and then returned to the jobsite for review and evaluation.

How Do You Evaluate Progress?

Progress should be measured against the activity benchmark. As illustrated earlier in this chapter, daily productivity can be very erratic depending on the work being performed that day, the physical conditions at the site, and the availability of the necessary people, materials, and equipment. Therefore, the important measure of progress is the average productivity and its trend. In the beginning of the work, time is required to perform layout, handle material, and set up for performing the work, which will result in a low productivity rate. However, as time goes on, the average productivity rate should continue to increase, resulting in a trend toward achieving the benchmark.

How Do You Communicate Progress?

Communicating progress to the crew is as important as communicating progress to the project manager and others at the home office. The benchmarking spreadsheet and run chart for each benchmarked activity should be prominently posted in the job trailer at the end of each day. The crew should be able to view the benchmarking information the next morning when crew members come in to start work. Posting this information every day lets the crew know how the crew is doing on a daily basis and will often result in an impromptu discussion about how the process could be improved and how to achieve better results. In addition, regular meetings should be scheduled by the project manager to review overall project progress as well as benchmarked activities to keep the foreman and crew informed of the project status, to identify current and potential problems, and to solicit suggestions for improving production processes.

How Do You Audit Production Data?

Auditing production data is a must to ensure that the benchmarking data provided by the crew is accurate. "Banking" work is a common practice when the crew does not trust the field benchmarking process or how the data will be used. Banking is sometimes used to inflate initial unit production to cover up for the inherent low production rate that usually occurs at the beginning of the activity due to most of the work being preparatory. As the work picks up and the deficit resulting from the preparatory work is "paid back," the crew may reduce the productivity reported, anticipating the lower productivity that results as the activity is finished and requires cleanup, trimout, and startup. Banking productivity defeats the purpose of field benchmarking and gathering contemporaneous data because it makes getting a true picture of what is going on impossible. The project manager and foreman must impress on the crew that the actual time spent on the benchmarked activity and the actual work put in place on a daily basis must be reported.

To make sure that benchmarking data is accurate, a quick audit of the material delivered to the jobsite, currently in inventory at the jobsite, and shown to be installed on the benchmarking report should be made on a regular basis. Too much or too little material on the jobsite may indicate that the crew is practicing banking. If banking is found, the project manager should work with the foreman and crew to find out what their concerns are about benchmarking and find a way to eliminate those concerns so that an accurate picture of activity and project progress can be developed.

FIELD BENCHMARKING PROCEDURE

This section provides a generic procedure for implementing field benchmarking in a construction firm. This generic procedure is intended to provide a starting point for a contracting firm to develop its own field benchmarking procedure that is tailored to its specific needs and operations. The generic field benchmarking procedure can be included as part of the contracting firm's quality assurance manual or operations manual.

Quality Assurance Procedure

Scope

This procedure applies to all construction projects except as determined by the vice president of operations.

Purpose

The purpose of this procedure is to provide a systematic and timely method for planning, monitoring, documenting, and controlling work in the field. Field benchmarking will result in greater employee involvement, more efficient construction operations, increased customer satisfaction, and sustained competitive advantage. More efficient construction operations result from increased jobsite safety; improved construction quality and reduced rework; and increased construction productivity through preplanning, continuous improvement, and process reengineering.

Personnel and Responsibilities

This section identifies key personnel and describes their role and responsibilities as related to field benchmarking.

The **vice president of operations** is responsible for the following:

1. Ensuring that benchmarking is performed on all construction projects in accordance with the scope section of this procedure.
2. Ensuring that all division personnel are trained in benchmarking

3. Ensuring that there are sufficient data processing resources to support field-benchmarking activities.
4. Working to improve the effectiveness and efficiency of the field benchmarking procedure.

The **project manager** is responsible for the following:

1. Preparing and submitting the project benchmarking plan to the accounting department prior to the start of construction. The project benchmarking plan must be in compliance with this procedure.
2. Setting project-specific benchmarking requirements for each project and communicating those requirements to the foreman.
3. Implementing the project benchmarking plan at the project site.
4. Working with the project foreman to ensure that all construction team members at the project site understand and are trained in benchmarking procedures.
5. Ensuring that benchmarking data from the project is received, processed, and distributed in accordance with the procedure and the project benchmarking plan for the project.
6. Working with the project foreman to ensure that reported benchmarking data is accurate and communicated to the construction team.
7. Ensuring that the benchmarking information is used for construction process improvements that will lead to greater safety, quality, and productivity at the project site.
8. Working to improve the efficiency and effectiveness of the field benchmarking procedure.

The **project foreman** is responsible for the following:

1. Working with the project manager to implement benchmarking on the project in accordance with this procedure and the project benchmarking plan.
2. Identifying the benchmarking training requirements of individual construction team members and notifying the project manager of those requirements.
3. Working with the project manager and the construction team to set rational and achievable benchmarks for those activities to be benchmarked based on actual field conditions.

4. Ensuring that construction team members gather the required production data accurately and turn in production data on a daily basis for analysis.
5. Ensuring that the production data gathered is used for construction process improvement only.
6. Sharing the production information with the construction team in a timely manner so that construction process improvements can be made as needed.
7. Conducting preplanning and brainstorming sessions with the construction team to improve construction process safety, quality, and productivity.
8. Working to improve the efficiency and effectiveness of the field benchmarking procedure.

Construction team members are responsible for the following:

1. Working with the project foreman to implement benchmarking on the project in accordance with this procedure and the project benchmarking plan.
2. Actively taking part in benchmarking training.
3. Actively participating in setting benchmarks.
4. Actively participating in brainstorming and preplanning sessions for process improvement.
5. Executing the work in accordance with the activity preplan
6. Gathering accurate personal production data on a daily basis and submitting it to the project foreman for aggregation and analysis.
7. Assisting the project foreman by reporting possible problems, anticipated material and equipment shortages, needed or broken production equipment, and providing suggestions for construction process improvement.
8. Working to improve the efficiency and effectiveness of the field benchmarking procedure.

The **data processing specialist** is responsible for the following:

1. Receiving benchmarking data from the field in accordance with this procedure and the project benchmarking plan.
2. Entering benchmarking data, processing benchmarking data, and outputting benchmarking information in accordance with this procedure and the project benchmarking plan.

Table 2.1 Reference Documents

Document Number	Document Title
001	Activity Preplanning Form
002	Labor Activity Code Index
003	Daily Activity Benchmarking Report
004	Foreman's Daily Project Log
005	Glavinich, T. E. (2008). "Benchmarking Field Operations." In *Construction Productivity* (Editor: Rojas, E. M.), *Strategic Issues in Construction Series*; Ft. Lauderdale, FL: J. Ross Publishing.

3. Distributing benchmarking information in accordance with this procedure and the project benchmarking plan.
4. Alerting the vice president of operations when it is anticipated that there will be insufficient resources available to meet the data processing demands of the field.
5. Working to improve the efficiency and effectiveness of the field benchmarking procedure.

Reference Documents

The documents shown in Table 2.1 are referenced in this procedure.

Definitions

All terms used in this procedure are intended to be defined in accordance with normal industry usage unless otherwise noted in this section. Terms defined in this section are unique to this procedure and the definitions provided or referenced apply.

Step-by-Step Requirements

This section provides the detailed step-by-step requirements for implementing field benchmarking (see Figure 2.5). Responsibility for performing each step should be defined by the approved project benchmarking plan unless responsibility is specifically assigned in this section.

1.0 Develop Project Benchmarking Plan

1.1 Review contract documents.

1.2 Review bid estimate work breakdown and labor hours.

1.3 Identify activities to be benchmarked. Those activities that are identified as being critical to project success or high risk, are repetitive

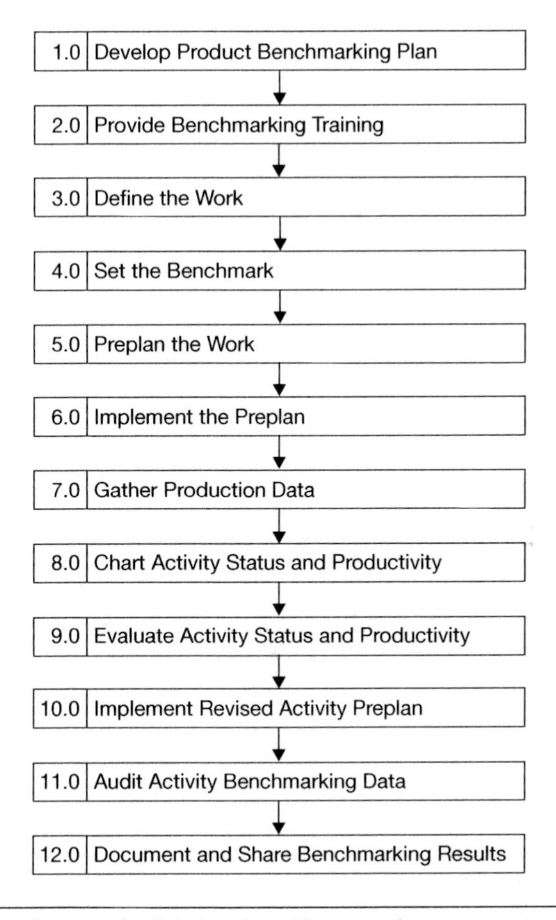

Figure 2.5 Summary flowchart for field benchmarking procedure.

and contain a significant number of labor hours, and/or are found to be a potential problem are to be benchmarked. Change order work should be benchmarked separately from base contract work. Specific instruction on how to identify activities to be benchmarked can be found earlier in the chapter.

1.4 Determine benchmarking data-gathering and processing requirements.

1.5 Determine benchmarking report frequency and format requirements.

1.6 Assign benchmarking data-gathering, processing, and reporting responsibilities.

1.7 Prepare a written project benchmarking plan that addresses Items 1.1 through 1.6 as well as any other items that are important to ensure a successful benchmarking effort. The project benchmarking plan must be prepared in accordance with this procedure.

1.8 Project manager submits the project benchmarking plan to the accounting department.

1.9 Implement the project benchmarking plan.

2.0 Provide Benchmarking Training

2.1 Identify training required to successfully implement the project benchmarking plan.

2.2 Develop a benchmarking training plan and schedule.

2.3 Provide needed benchmarking training.

3.0 Define the Work

3.1 Define the scope of work to be benchmarked, which includes a description of the work activity and all work involved in accomplishing that work activity. Specific instructions on how to define the scope of work to be benchmarked are provided earlier in the chapter.

3.2 Assign a labor code to the work activity in accordance with the labor activity code index. Modify standard description as required to match project requirements. Assign any additional labor codes required in accordance with the company procedure.

3.3 Document activity labor code and scope of work to be benchmarked.

3.4 Provide work activity information to the project foreman and data processing specialist.

4.0 Set the Benchmark

4.1 Prior to starting work on a work activity identified for benchmarking, meet with the construction team who will perform the work to set the benchmark production rate in labor hours per unit. The benchmark production rate unit shall be the same as that used to prepare the bid estimate in order to facilitate comparison and cross-reference.

4.2 Share activity scope of work, bid estimate labor hours, and other relevant information with the construction team who will perform the work.

4.3 Consider bid estimate labor hours, work unit quantity, current and anticipated site conditions, available means and methods, and other relevant information in setting the activity benchmark.

4.4 Set the benchmark productivity rate for the work activity.

5.0 Preplan the Work

5.1 After setting the benchmark for an activity and before starting work on the activity, prepare an activity preplan.

5.2 Meet with the construction team who will perform the work to determine the means and methods for performing the work that will achieve the benchmark. Use the activity preplanning form as a guide for planning the execution of the work.

5.3 Brainstorm work process improvement methods for improving safety, quality, and productivity with the construction team and incorporate them into the activity plan.

6.0 Implement the Preplan

6.1 Ensure that the needed information, materials and installed equipment, tools and production equipment, and space are available in sufficient quantities when needed by the construction team.

6.2 Commence work in accordance with the activity preplan and project schedule.

7.0 Gather Production Data

7.1 Gather daily activity production data from each construction team member. In addition, gather construction team comments with regard to what work was performed during that day, problems and delays encountered, and suggested process improvements. If no work was performed on the work activity on a given day, that should be noted as well, along with the reason that no work was performed.

7.2 Sum the daily activity production data from each construction team member to get the total daily production data for the work activity.

7.3 Record the total daily construction team production data for the activity on the daily benchmarking activity report form. The daily benchmarking activity report form may be either the standard paper form or an electronic spreadsheet version that provides the same information in the same format.

7.4 Record construction team comments with regard to what work was performed during the day, problems and delays encountered, and suggested process improvements in the foreman's daily project log and/or on the daily benchmarking activity report.

8.0 Chart Activity Status and Productivity

8.1 Process production data in accordance with this procedure and the project benchmarking plan.

8.2 Prepare a benchmarking summary spreadsheet that includes the following work activity information and data columns:
   8.2.1 Work Activity Information:
      (a) Activity Description
      (b) Activity Labor Code
      (c) Activity Scope
      (d) Estimated Production Rate
      (e) Benchmark Production Rate
   8.2.2 Work Activity Data Columns:
      (a) Date
      (b) Effort:
         (1) Daily (or Weekly) Labor Hours Expended
         (2) Total Labor Hours Expended to Date
      (c) Installed Quantity:
         (1) Daily (or Weekly) Units Installed
         (2) Total Units Installed to Date
      (d) Unit Productivity:
         (1) Daily (or Weekly) Labor Hours per Unit
         (2) Average Labor Hours per Unit to Date
   8.2.3 Detailed instructions for setting up a benchmarking summary sheet and the associated column calculations are provided earlier in the chapter.

8.3 Provide a benchmarking summary run chart that plots daily labor hours per unit and average labor hours per unit taken from the benchmarking summary spreadsheet against time in days or weeks. Detailed instructions for setting up a benchmarking summary run chart are provided earlier in the chapter.

8.4 Distribute the benchmarking summary spreadsheet and run chart in accordance with the project benchmarking plan.

9.0 Evaluate Activity Status and Productivity

9.1 Evaluate the effectiveness of the activity preplan by comparing the average labor units to date against the benchmark productivity rate. Consider trends, type of work performed to date, problems and delays encountered, and suggested process improvements.

9.2 Post and/or distribute benchmarking summary spreadsheet and run chart for construction team review.

9.3 Meet with construction team in accordance with the project benchmarking plan and analyze activity progress, productivity, problems and delays encountered, and suggested process improvements.

9.4 Revise activity preplan as required to improve safety, quality, and productivity based on actual field conditions encountered.

10.0 Implement Revised Activity Preplan

11.0 Audit Activity Benchmarking Data

11.1 Audit activity benchmarking data on a periodic basis as required by the project benchmarking plan to ensure accurate data.

11.1.1 Compare total labor hours expended to date against job cost report labor hours.

11.1.2 Compare total units installed to date against actual material delivered less jobsite inventory.

11.2 Correct and document any errors identified.

11.3 Issue a revised benchmarking summary spreadsheet and run chart if needed.

12.0 Document and Share Benchmarking Results

12.1 Integrate activity benchmarking data into the company planning and estimating database.

12.2 Share innovations and construction process improvements throughout the company through in-house meetings and newsletter.

LEARNING THROUGH FIELD BENCHMARKING

Field benchmarking provides a unique opportunity for organizational learning and growth. Learning and continuous process improvements play an important role in maintaining or gaining competitive advantage in the marketplace. This section discusses how to use field benchmarking to improve construction processes and a contracting firm's bottom line.

Communicating Process Improvements

Process improvements that result from field benchmarking need to be documented and communicated company-wide. Current and future projects can benefit from what is learned on a particular project, but profiting from experience on a particular project can only happen if what is learned is communicated. Often, estimators, foremen, project managers, and others are faced with similar situations that have been addressed either successfully or unsuccessfully by others on other current and past projects. If the knowledge and experience gained is not communicated, then other projects will need to reinvent the wheel. This is a costly and inefficient way of doing business for the contracting firm and a waste of resources.

Communicating Continuous Process Improvements

Process improvements can be communicated within the contracting firm in many ways. However, first, upper management must create an environment that fosters interaction and actively supports the open exchange of information and experience throughout the company. People at all levels should be encouraged to share their knowledge and experience through informal and formal interaction. Additionally, management must promote experimentation and calculated risk taking and show that it understands that innovative ideas do not always work as planned. Learning from mistakes can be more valuable than learning from successes, especially when a particular mistake may be repeated numerous times on numerous projects.

Structured methods that can be used to communicate experience gained through field benchmarking include the following:

- Project cross-pollination
- Project case studies
- Continuous improvement database
- Continuous improvement forum
- Continuous improvement newsletter
- Field benchmarking electronic bulletin board

Project Cross-Pollination

Project cross pollination involves assigning project personnel to projects based on the project needs and their personal experience. For instance, if a foreman has experience installing a particular system or material that is critical to a project's success, that foreman would be assigned to that project in order for the project to benefit from his or her experience. Bringing in the experienced foreman will help avoid mistakes that the project team might have made without previous experience, as well as provide an opportunity for the project team to learn from the foreman. At the end of the project, the contracting firm has expanded its expertise with that particular system or material and more people know how to perform the work, increasing the firm's flexibility in assigning people to the next project that employs the same or a similar system or material. In addition, the project team working with the experienced foreman may be able to improve the existing process further, which may not be possible if the project team is starting from scratch without the experienced foreman.

Project Case Studies

Written project case studies can be developed by the project team to describe a particular process improvement that they learned from field benchmarking on a particular project. It is best if these case studies are drafted and submitted by the project team rather than just the project manager or foreman. If the project team drafts the case study of the process improvement, the case study will address the process improvement from a number of viewpoints, which will be helpful when another project manager or foreman applies it to another project.

Once written, the team could present the case study at a continuous-improvement forum or publish it in a newsletter. The case study could also be included in a searchable database of best practices that is either paper-based or electronic. The database allows estimators, foremen, project managers, and others to find information about how a particular issue was handled in the past as well as the success of the approach. Most importantly, it will identify expertise within the firm, and the people with that expertise can answer specific questions for the project team, provide additional and more in-depth information about a particular aspect of the installation or operation, and possibly serve as an internal consultant on the project as required.

Due to the time and effort involved with putting together project case studies, those who develop them should be recognized and rewarded for their efforts. In addition, a formal way of disseminating the information that permits feedback from others within the contracting firm should be devised and implemented. Ways of addressing the need for dissemination and peer review of project case studies include developing a continuous process improvement database, implementing a continuous process improvement forum, and recognizing innovative ideas in a newsletter.

Continuous Improvement Database

As noted, the continuous improvement database is a repository of best practices developed within the contracting firm and archived in either paper format, electronic format, or both. In paper format, the best practice database could be provided to project engineers, estimators, project teams, and others as an indexed loose-leaf notebook that could be easily updated and expanded over time. In electronic format, the database could be stored on the firm's server and accessed either in the office or remotely through the firm's intranet. Either system will work and the choice will depend on the contracting firm's information technology capabilities and user preferences.

Continuous Improvement Forum

Regular meetings can be scheduled by the contracting firm to disseminate information about what is learned on projects as a result of field benchmarking. At the meetings, case studies can be presented and best practices discussed. In addition, the mechanics of field benchmarking and its application in the field can be reviewed to improve this process as well. The continuous improvement forum provides a regular opportunity for estimators, project managers, foremen, and others to get together and discuss how field operations can be improved in order to increase safety, quality, and productivity as well as lower costs.

Continuous Improvement Newsletter

The continuous improvement newsletter can be a standalone internal newsletter that discusses both field and home office process improvements or it can be a section of the contracting firm's existing general newsletter dedicated to process improvement in the field. Articles addressing field benchmarking and construction process improvements would not replace the detailed project case studies but instead provide a summary of the project and construction process improvements and direct the reader to the project case study for more information. As with the continuous improvement database, the continuous improvement newsletter can be paper-based, electronic, or both. The goal of the continuous improvement newsletter is to make everyone in the contracting firm aware of its continuous process improvement program, keep employees and customers excited about the results of the initiative, and build a culture that values information sharing at all levels and a dedication to excellence.

Field Benchmarking Electronic Bulletin Board

An electronic bulletin board that can be accessed by estimators, project managers, foremen, and other contracting firm personnel can provide a valuable adjunct to the project case studies and their dissemination. The electronic bulleting board allows personnel to post questions about field operations and field benchmarking and get answers from others who are plugged into the electronic bulletin board. This would be a quick and easy way to get information quickly and easily and tap into the company's collective knowledge and experience. Also, because the bulletin board is electronic, it can be accessed and questions can be answered from anywhere, including the home office and project jobsites. For instance, if an estimator is putting together a bid estimate for a potential project but he or she is not sure how to estimate a particular piece of the project, the estimator could post a question on the electronic bulletin board. Project managers, foremen, and others who may have encountered this situation or something similar before can respond to the question. Based on the responses, the estimator not only has additional information but also a group of knowledgeable individuals who he or she can contact with more detailed questions. The electronic bulletin board will save contracting firm personnel a great deal of time, help the organization avoid repeating past mistakes over again, and build innovative processes discovered through field benchmarking into bid estimates, making the firm more competitive.

Support of Change Orders and Claims

Field benchmarking can be used to support change order requests and claims if the data provided is accurate and audited. As can be seen from Figure 2.3, the daily productivity varies considerably, while the average productivity demonstrates a fairly constant downward trend. The daily productivity data is contemporaneous and can be valuable in the support of change order requests and claims for lost productivity and disruption. The daily productivity data usually complements the daily report narrative and supports days and periods of poor productivity. Having the daily productivity data available on key project activities can save the contracting firm a great deal of time and effort in the preparation of change order requests and claims. In addition, the data that the change order request or claim is based on is not derived from aggregate project cost and labor information but on actual daily activity recorded at the time the work was being performed.

Closing the Loop with Estimating

Field benchmarking also provides an excellent opportunity to close the loop with estimating. Worker-hour and productivity data gathered under actual field conditions can be used to analyze the variances between the amount of labor bid and what was actually used. The impact of actual field conditions can also be factored into the analysis to determine what adjustments could be made for similar installations and projects in the future. Using field benchmarking data to close the loop with estimating can significantly improve the accuracy of estimates and project profitability.

CONCLUSIONS

The purpose of this chapter was to promote the use of field benchmarking to improve the quality, productivity, and safety of construction. Field benchmarking increases worker participation in the decisions that affect the daily work and the quality of the workplace environment. Effective field benchmarking helps firms to develop mutually agreed-to performance goals, design work processes to achieve these goals, and set performance criteria for measuring the effectiveness of work processes. Benchmarking can be a key element of a contracting firm's continuous-improvement program. Benchmarking provides the yardstick against which improvement efforts can be measured. Without measurement, there is no way of knowing if progress is being made toward goals or not. Field benchmarking is a special case of benchmarking that focuses on improving the contracting firm's field operations.

A word of caution: Field benchmarking is not about evaluating crew or individual worker performance. Field benchmarking is about evaluating and improving the work processes that the crew works within. For field benchmarking to work, the contractor must take responsibility for many of the problems in the field, listen to and value worker's suggestions about how to improve the process, and work directly with the worker to improve the process. Field benchmarking requires trust and openness on both sides, which in turn requires an investment of time, effort, and training.

WEB ADDED VALUE

Procedure.doc

This file provides the procedure for implementing field benchmarking in a construction firm included in the "Field Benchmarking Procedure" section of this chapter. This generic procedure is intended to provide a starting point for a contracting firm to develop its own field benchmarking procedure that is tailored to its specific needs and operations. The digital version of this generic field benchmarking procedure can be easily incorporated as part of the contracting firm's quality assurance manual or operations manual.

This book has free material available for download from the Web Added Value™ resource center at *www.jrosspub.com*

3

STACKING OF TRADES

Dr. Awad S. Hanna, *University of Wisconsin-Madison*
Dr. Jeffrey S. Russell, *University of Wisconsin-Madison*
Erik O. Emerson, *University of Wisconsin-Madison*

INTRODUCTION

Labor productivity is one of the most important factors that affect the profitability of contractors. Stacking of trades is one of the many influences that affect labor productivity. This chapter focuses on the qualitative and quantitative aspects of stacking of trades in order to help contractors create useful productivity monitoring systems.

Stacking of trades is defined as operations that take place within physically limited space with other contractors. Stacking of trades relates the number of different trades (pipefitters, electricians, etc.) within a measured work area to labor productivity. Stacking of trades should not be confused with other concepts such as overcrowding and overmanning. Overcrowding is defined as the increase of all trade types within a given construction work area, while overmanning relates the change in the size of a particular trade to the change in the trade's productivity.

When contractors prepare a bid or a change order estimate, they must make reasonable assumptions. One of the basic assumptions is that the contractor's workforce will be able to perform its work with minimal interference from other trades. However, in many instances, contractors do not plan for the influence that stacking of trades can have on labor production and are not compensated appropriately when stacking of trades develops on a worksite.

Several options are available to deal with the effects that stacking of trades has on labor productivity. Contractors may be able to change construction methods, work sequence, crew composition, or the hours of the day when work is to be performed. By understanding the causes of stacking of trades, and creating a methodology to measure its effects on labor productivity, contractors can create more accurate estimates. This, in turn, allows for proper compensation, increase accuracy in scheduling, and an overall improvement of the construction process.

PREVIOUS STUDIES ON OVERCROWDING, OVERMANNING, AND SITE DENSITY

As mentioned, overcrowding is the increase of all labor types within a given construction work area. Overcrowding uses the percent increase of all trades, without specifying which crafts are within the work area. Overcrowding is frequently mentioned in productivity literature, and many papers make reference to a model produced by the U.S. Army Corps of Engineers (1979). This model shows a relationship between overcrowding and percent change of productivity. In the model, percent overcrowding is calculated by taking the total increase of the number of workers in all trades within a work area and dividing it by the previous total. Loss of productivity, in percentage terms, is expressed as a drop of actual productivity from a theoretical maximum or estimated capacity of the crew. The major problem with this model is that there is no supporting data. There is no background information on how the model was developed, or any statistical analysis. Without this information, it is difficult to know when to use the model or under what circumstances the model was developed. Furthermore, the accuracy of the model is unknown.

The second problem with this model, and all overmanning models, is the use of the percent increase of crew size as the independent variable. This independent variable can penalize smaller initial work crews significantly more than larger initial work crews. For example, let's suppose that two equally sized work areas, one with three workers and the other with nine workers, had one worker added to each crew. The percent increase method would indicate that the smaller crew of three would be 33% overcrowded, while the other crew of nine workers would be 11% overcrowded, implying that the smaller crew would be significantly less productive than the larger crew when crew size is increased by the same amount.

Finally, this method does not consider the size of the work area. A small work area can experience significant productivity losses if the crew size is only slightly increased. A large work area may need a significantly larger increase of crew size in order to experience any productivity loss. If all other variables except the size

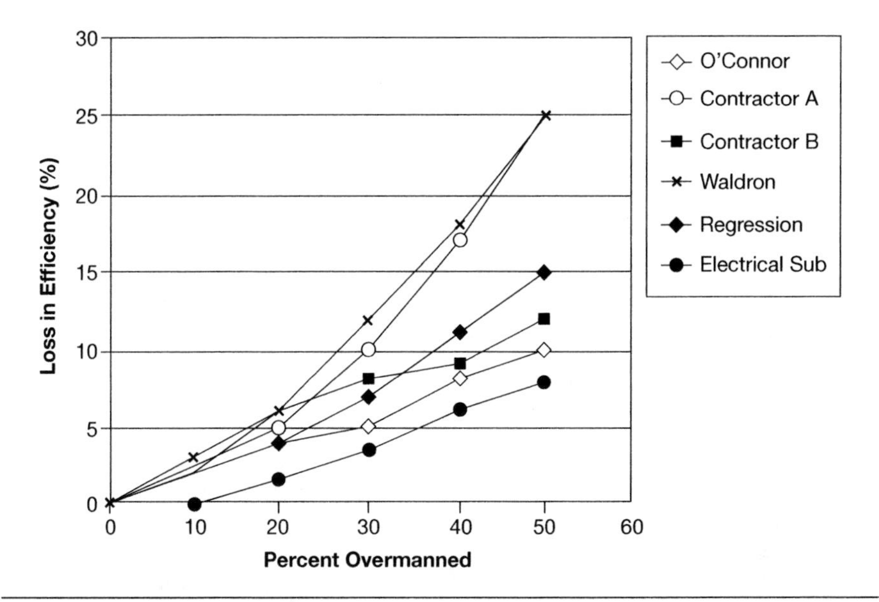

Figure 3.1 Thomas Literature Review.

of work area were constant, the percent increase method would indicate that productivity would be the same in a small and large work area.

Overmanning, as mentioned, is the relationship between the percentage change in the number of workers of a particular trade and its productivity. Overmanning is based on the law of diminishing returns. This law states that as each worker is added to a crew, the new worker is not as productive as the previously added worker. Eventually, the addition of a worker will drive down the productivity of the entire crew. The law of diminishing returns is well documented in economic sources.

Thomas (1994) provides a review of overmanning models. This review, shown in Figure 3.1, uses percent drop from an expected or planned productivity to measure loss in efficiency. The models also use percentage overmanned (increase) as the independent variable.

Site density compares size of work area and number of workers within the work area to the site's overall productivity. This method has two major advantages over the percent increase methods of overcrowding and overmanning. First, this method is not based on previous crew size or change in crew size; therefore, it does not penalize smaller work crews. Second, this method takes into account the size of the work area and therefore does not treat different-sized work areas the same.

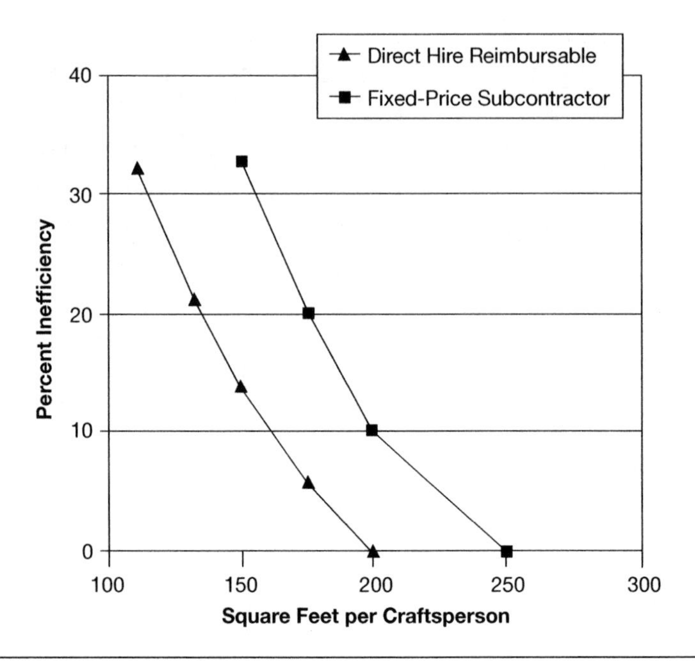

Figure 3.2 Mobil Oil site density model.

Figure 3.2 shows the results of a study performed by the Mobil Oil Company. This study uses site density as the independent variable. Site density is calculated by dividing the measured work area by the number of workers within a work area. This model uses inefficiency as the dependent variable. Inefficiency is the percentage of productive time lost or drop of productivity from a fixed theoretical value or estimated rate of production.

Figure 3.3 shows the results of a study on-site density performed by Smith (1987). This study also uses site density as the independent variable. Site density is calculated using the same technique as the Mobil study; however, Smith's study uses squared meters to measure the size of a work area instead of squared feet. Figure 3.3 directly measures productivity on the vertical axis. Productivity is defined as actual production divided by a fixed theoretical value or estimated rate of production. Smith (1987) states that "on a reasonable densely populated site, 30 m^2 per person, each increase in labor force by 50% will reduce productivity 10%."

These two site density models suffer from similar deficiencies. First, there is little background information or data presented with the models. Second, there are no statistical procedures mentioned for the generation of the models. Third, they do not take into account the special needs of different trades, different work performed by certain trades, and the general lack of trade interaction. These shortcomings make it difficult to use these relationships effectively.

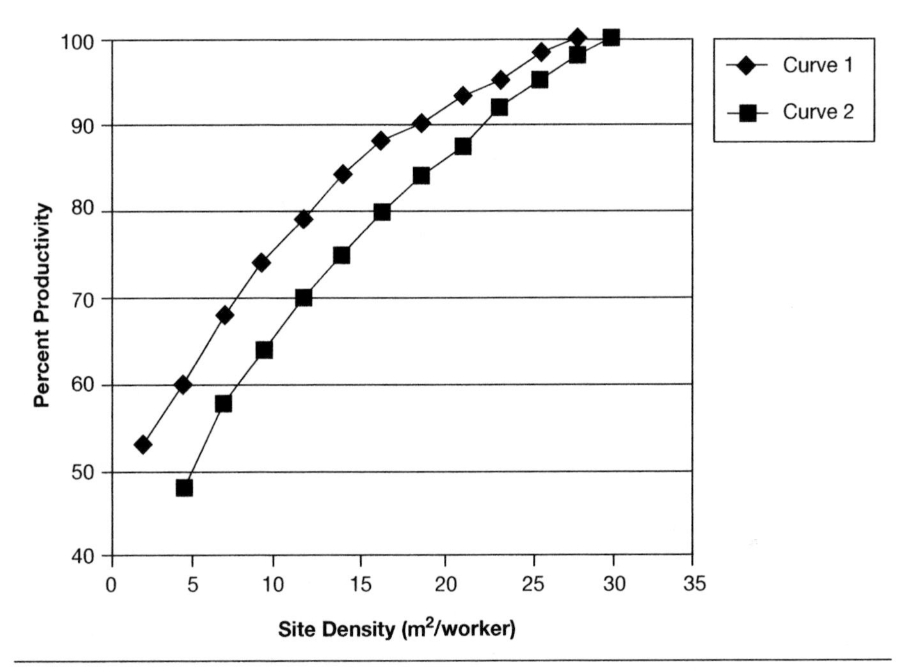

Figure 3.3 Smith site density model.

UNDERSTANDING STACKING OF TRADES ISSUES

To better understand the issues related to stacking of trades, a qualitative survey was developed. The survey was randomly distributed to 500 members of NECA and had a response rate of 10%, typical for this type of study. The survey collected demographic data and opinions specifically on stacking of trades.

Demographics

The survey respondents were 86% upper managers (president/owner or vice presidents), 8% project managers, and 6% in other positions. Contractors from across the United States responded to the survey, with 45% of the responses from the Midwest, 25% from the West Coast, 15% from the South, and 15% from the East Coast.

The commercial and industrial sectors accounted for 32% of the responses, the manufacturing sector generated 13%, both the outside/line and residential sectors each generated 7%, and the remaining responses (9%) were in the other category.

Companies with a five-year sales average below $1 million generated 16% of the responses, 36% were generated from contractors whose five-year sales average was between $1 million and $5 million, and the rest (48%) were from contractors that made more than $5 million on average during the last five years.

Out of the contractors surveyed, 84% had 175 or fewer construction workers. The majority of contractors surveyed, 69%, had been in existence for 50 or fewer years, with the largest response coming from contractors that have been in existence for 31 to 40 years (22%). The majority of contractors, 57%, spent somewhere between 61% to 100% of their time as subcontractors.

Qualitative Data

The survey collected the expert opinion of respondents on a variety of issues related to stacking of trades. All results were analyzed using two-way ANOVA and deemed to be significant at the 95% confidence interval.

Pre-Bid Indicators

Contractors were asked to rank nine indicators that a project was going to experience problems with stacking of trades before bidding on the project: (1) knowing the prime contractor, (2) knowing the owner, (3) knowing the designer, (4) type of work performed, (5) location of work, (6) prime contractor's scheduling practice, (7) your overall company's work schedule, (8) project delivery method, and (9) site condition.

Contractors were allowed to rank each indicator with a number from 1 to 5. The number 1 would denote a strong relationship between stacking of trades and the pre-bid indicator. A ranking of 5 would denote a weak or nonexistent relationship between stacking of trades and the pre-bid indicator. The averages of pre-bid indicators are shown in Table 3.1. From this analysis, prime contractor's scheduling practices and the type of work to be performed came on top as key indicators that a project is going to have difficulties with stacking of trades.

Causes of Stacking of Trades

Contractors were asked to identify the causes of stacking of trades on a construction site. They were given seven possible causes to rank using the same technique as in the pre-bid indicators: (1) rework, (2) scope change, (3) change order, (4) project acceleration, (5) complexity of work, (6) poor planning, and (7) delay in preceding activity.

The averages of this analysis are shown in Table 3.2. From the analysis, respondents identified a strong correlation between project acceleration and delay in previous activity and stacking of trades. Change orders, scope changes, and poor planning showed a weaker correlation but may still be causes of stacking of trades.

Table 3.1 Average Scores for Pre-Bid Indicators

Pre-Bid Indicators	Rank	Average Score
Prime Contractor's Scheduling Procedure	1	1.40
Type of Work Performed	2	1.53
Site Conditions	3	2.20
Knowing the Prime Contractor	4	2.60
Location of the Work	5	2.73
Project Delivery Method	6	3.07
Contractor's Work Schedule	7	3.47
Knowing the Owner	8	3.47
Knowing the Designer	9	3.87

Table 3.2 Average Scores for Causes

Cause	Rank	Average Score
Project Acceleration	1	1.13
Project Delay	2	1.52
Change Orders	3	1.84
Poor Planning	4	1.87
Change on Scope of Work	5	2.03
Complexity of Work	6	2.87
Rework	7	2.97

Timing of Stacking of Trades

Contractors were asked to assign a percentage for stacking of trades to each of the four quartiles of a project's duration according to the preponderance of occurrence. The larger the percentage, the more likely that stacking of trades would develop during the given quartile of a project's duration. Quartile 1 represents the first 25% of the project's duration, quartile 2 represents the second 25% of the project's duration, and so on.

The averages of the percentages assigned to each quartile are shown in Table 3.3. The average for each quartile tend to increase as the project advances, with the fourth quartile having a significantly higher value than all others. This seems to indicate that, in the opinion of the respondents, there is preponderance in the occurrence of stacking of trades toward the end of projects.

Table 3.3 Average Percentage Values for Project Quartiles

Quartile of Project Duration	Percentage of Occurrence
1	10.5
2	14.9
3	27.9
4	49.3

Table 3.4 Average Scores for Effects

Aspect of Work	Rank	Average Score
Limitations of Work Area	1	1.41
Work Sequencing	2	1.67
Crew Coordination	3	1.85
Scheduling	4	1.95
Material Handling	5	2.79

Effects of Stacking of Trades

Contractors were asked to rank, on the 1 to 5 scale previously described, how stacking of trades affects different aspects of their work. They were given five areas of work: (1) material handling, (2) work sequencing, (3) limitations of work area, (4) crew coordination, and (5) scheduling.

The averages of the scores are shown in Table 3.4. From the analysis, stacking of trades seems to impact all aspects of work, with material handling being the least affected area.

Detecting Stacking of Trades

Contractors were asked if they had a productivity monitoring system that could measure or detect the affects of stacking of trades. To accomplish this, the survey asked two questions. The first question addressed whether the contractor had an "early warning system" that would alert them that a project was going to have problems with stacking of trades. Some of the respondents, 42%, indicated that they had an early warning system. When ask to expand on their early warning system, the majority of the responses indicated the use of expert opinion and basic project management tools.

The next question asked was whether the contractor's productivity monitoring system detected the effect of stacking of trades. The majority of the survey's

Table 3.5 Average Scores for Rationale for Compensation

Justification for Compensation	Rank	Average Score
Performing Work in Restricted Workspaces	1	1.43
Increases in the Amount of Manpower Needed	2	1.51
Performing Work out of the Planned Sequence	3	1.57
Increases in the Amount of Crew Idle and Waiting Time	4	1.66
Increases in the Amount of Overtime Needed	5	2.14
Increases in the Amount of Time Monitoring Needed	6	2.54
Increases in the Amount of Equipment Needed	7	2.69
Performing Rework	8	2.80
Increases in Material Handling	9	2.97

responses, 72%, indicated that the current systems of measuring productivity did not detect the influence of stacking of trades.

Rationale for Compensation

When stacking of trades develops on a site, a contractor can experience losses in production. Contractors were asked to rank nine reasons that justify compensation when stacking of trades is present. The same ranking system previously described was used. The reasons for justifying compensation included the following: (1) increase time when the work crew is idle, (2) need for more equipment, (3) working out of sequence, (4) rework, (5) increase in the amount of manpower, (6) use of overtime, (7) increase material requirements, (8) increase time spent monitoring work crews, and (9) working in constricted workspaces.

The averages scores are shown in Table 3.5. From the analysis, contractors believe that they should be compensated for stacking of trades mostly because of having to work in restricted workspaces, increases in the amount of manpower used, working out of sequence, and increases in the amount of crew idle and waiting time.

Summary

The primary purpose of the survey was to gather knowledge about the qualitative aspects of stacking of trades. The following conclusions can be made from the survey responses:

1. The most relevant pre-bid indicators for stacking of trades are prime contractor's scheduling procedures and the type of work performed.

2. Project acceleration and delay in previous activity are seen as the most likely causes of stacking of trades. Change orders, scope changes, and poor planning may also cause stacking of trades.
3. There seems to be a preponderance of stacking of trades toward the end of projects. The third and fourth quartiles of a project's duration are more likely to have difficulties with stacking of trades than the first and second quartiles of a project's duration.
4. Stacking of trades seems to impact all aspects of work, with the least impact being on material handling.
5. Most contractors lack the capacity to accurately detect the impact of stacking of trades on labor productivity.
6. Contractors believe that they should be compensated for stacking of trades mainly because of working in restricted workspaces, increases in the amount of manpower used, working out of planned work sequence, and increases in the amount of crew idle and waiting time.

QUANTITATIVE ANALYSIS

After the collection of qualitative data, our study concentrated on the collection and analysis of quantitative data. The objective of this effort was to develop a quantitative model to explain loss in efficiency due to stacking of trades. Figure 3.4 is a graphical depiction of the procedure used in the quantitative analysis.

Raw Data Requirements and Raw Data Provided

A data collection form was developed to provide a standard tool for data collection. The data collection concentrated on collecting raw data on (1) specific work activity, (2) number of workers, (3) types of workers, (4) size of work area, (5) estimated work hours for an activity, and (6) actual work hours for an activity. A copy of the data collection sheet is shown in Figure 3.5. The data collected from contractors is shown in Table 3.6. Contractors provided weekly productivity reports, a copy of the project's plans, and a report on how many other trades were on the project. Many electrical contractors did not directly collect information on the number of nonelectrical workers on the project. Typically, the electrical contractor had to ask the prime contractor or the owner representative for information on the number of nonelectrical workers.

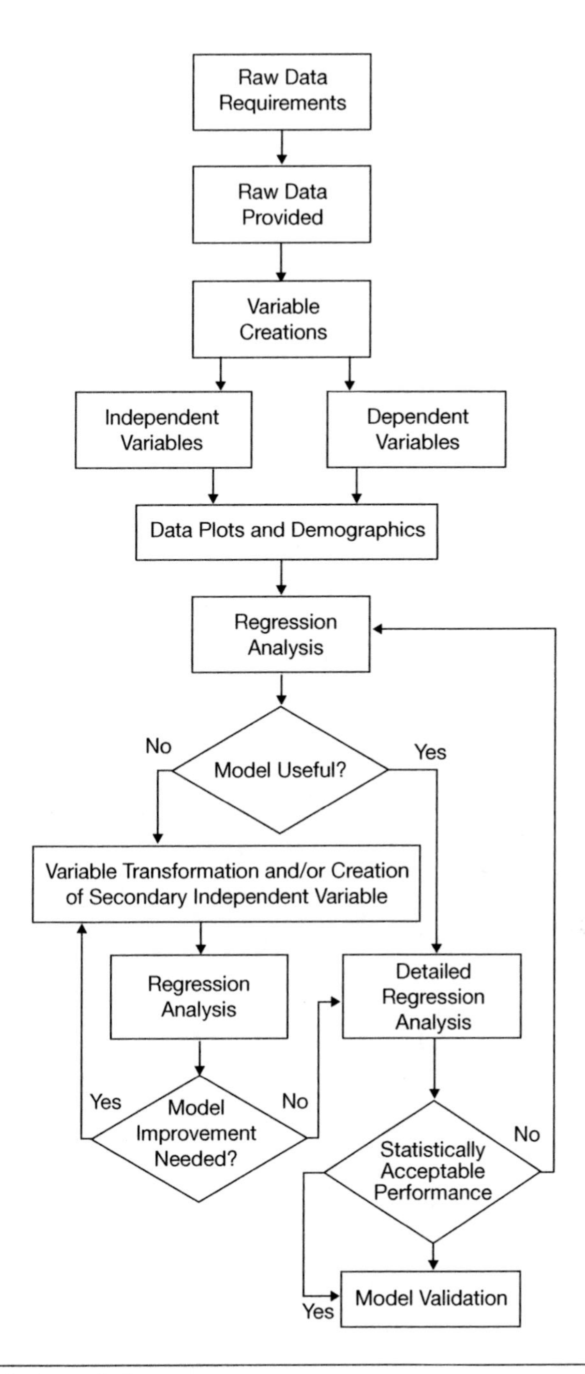

Figure 3.4 Quantitative analysis process.

Stacking of Trades Productivity Survey

The purpose of this form is to quantify the effects of stacking of trades and site congestion

Name:__________ Project Name: __________ Start Date: __________ Finish Date: __________

	Dimension (FT)**			Hours		Crew Size***				
Location*	Length	Width	Height	Estimated	Earned	Your Crew	Mechanical	Carpenter	General Contractor	Other

*****Location:** Portion or part of building or facility where Stacking of Trades occurs.
******Dimension:** Dimensions of area where Stacking of Trades occurs. All dimensions should be in feet.
*******Crew Size:** The number of individuals in each crew type that occupy the same location that you are working in.

Figure 3.5 Sample quantitative data collection sheet.

Table 3.6 Information Generated from Data Collected

Electrical Density (ft^2/worker)	Nonelectrical Density (ft^2/worker)	Efficiency (Actual Hours/Estimated Hours)
120.00	96.00	1.12
156.00	89.14	1.15
1,200.00	200.00	1.18
517.48	132.52	1.21
472.48	98.79	1.38
374.72	92.09	1.36
663.60	331.80	1.48
5,125.86	855.98	1.21
443.50	295.67	1.23
3,679.00	566.00	0.90
4,539.00	534.00	1.02
3,026.00	432.29	0.99

Factors Affecting Labor Efficiency (Independent Variables)

In our model, we define two main factors affecting labor productivity when stacking of trades develops on a site: electrical and nonelectrical density. As a result, the first step in the process of developing the statistical model was to measure work area. Work area, in square feet, is defined as the total area where electrical work is performed minus any unusable area.

Work area was used in determining electrical and nonelectrical density. Electrical density is calculated by dividing the work area by the number of electrical workers within that area. The nonelectrical worker density is calculated analogously by dividing the same work area by the number of nonelectrical workers within that area.

The dependent variable was defined as the ratio of actual labor-hours expended to estimated labor-hours or actual rate of productivity divided by estimated. We consider this ratio to be a proxy for labor productivity.

Demographics and Data Plots

Five separate electrical contractors provided seven projects that generated 12 data points. One contractor provided five points, while another contractor provided only 1 data point. On average, electrical workers represented 24.1% of the total labor force on the construction site. The data points were provided from the

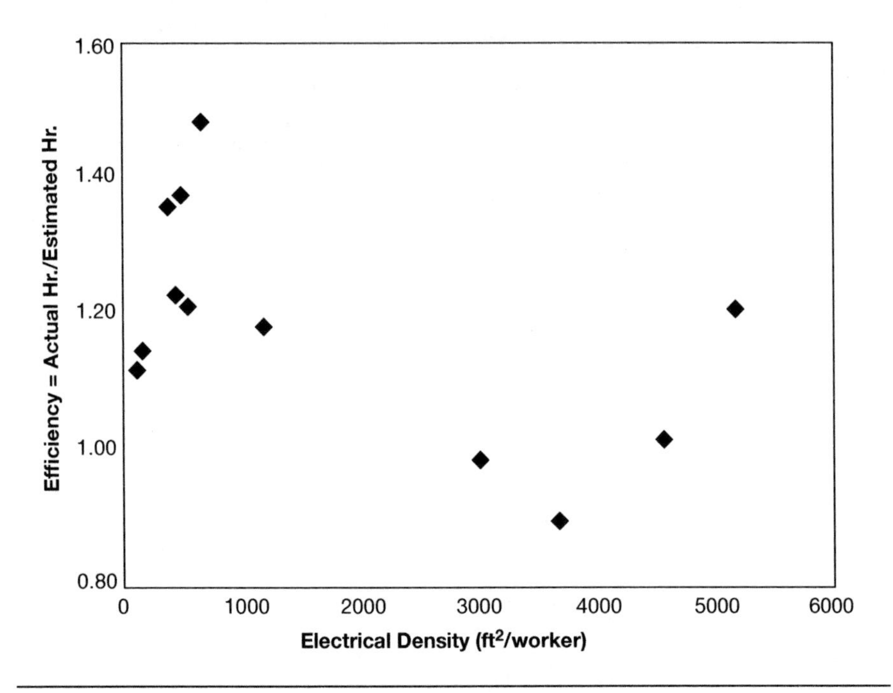

Figure 3.6 Scatter plot of (actual hours/estimated hours) to electrical density.

following types of projects: power plant remodeling, industrial remodeling, hospital construction, remodeling of a research/higher-education building, and office construction. The data were collected from electrical contractors performing work in the following types of work area: electrical service area, mechanical rooms, electrical rooms, computer processing center, office/laboratory space, and production space.

A scatter plot of the data is shown in Figure 3.6. This graph has electrical density on the horizontal axis expressed in squared feet by electrical worker. The vertical axis has labor efficiency represented as the ratio of actual electrical hours divided by estimated electrical hours. This graph suggests a curved relationship between electrical density and the ratio of actual electrical hours to estimated electrical hours.

Figure 3.7 shows the relationship between nonelectrical density and electrical labor productivity. This graph has nonelectrical density on the horizontal axis, expressed in squared feet by nonelectrical worker. The vertical axis has labor efficiency expressed as the ratio of actual electrical hours divided by estimated electrical hours. This scatter plot also indicates that there may be a curved relationship between nonelectrical density and the ratio of actual electrical hours to estimated electrical hours.

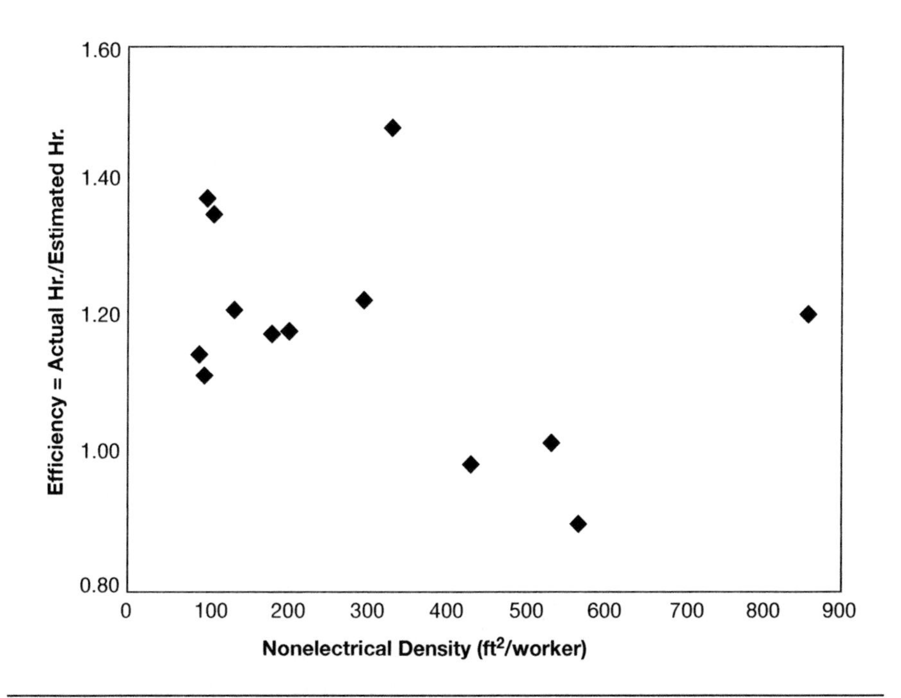

Figure 3.7 Scatter plot of (actual hours/estimated hours) to nonelectrical density.

Regression Analysis

There are several methods to generate statistical models. A two-step approach was used in this study. A simple regression analysis was performed to find promising models. The simple regression analysis generated the following statistical tests: R^2, adjusted R^2, and a Cp value. A more detailed regression analysis was then performed on the most promising models. The more detailed regression analysis included the following tests: F test, T test and residual analysis.

The initial attempts to develop a model using electrical and nonelectrical density did not create a model with adequate R^2, adjusted R^2, and a Cp value to warrant further investigation. The next step was to square electrical and nonelectrical densities and add a term that contained electrical density times nonelectrical density. Models containing these extra terms exhibited significant improvements in R^2 and adjusted R^2; however, the models had problems with the residual analysis. Transformation of the independent variables was the next operation performed in the regression analysis. Models with transformation of independent variables performed well, except for the residual analysis. The next transformation that was attempted was a power transformation. A power transformation is when both the independent and dependent variables are transformed using the natural logarithm. This is a standard statistical transformation and can be found in many

Table 3.7 Regression Analysis

Predictor	Coefficient	Standard Deviation	T Value	P Value
Constant	3.288	1.975	1.66	0.140
A	1.670	0.447	3.74	0.007
B	-3.173	1.094	-2.90	0.023
B^2	0.496	0.146	3.40	0.011
(A)(B)	-0.322	0.083	-3.89	0.006

Table 3.8 Analysis of Variance

Source	D.F.	S.S.	M.S.	F Value	P Value
Regression	4	0.172258	0.043065	5.45	0.026
Error	7	0.055287	0.00790		
Total	11	0.227545			

regression textbooks. This transformation generated an equation that had significant R^2 and adjusted R^2 and acceptable residual analysis. The equation generated by the regression analysis based on a logarithmic transformation is as follows:

$$P = 3.288 + 1.67(A) - 3.173(B) + 0.4962(B^2) - 0.3225[(A)(B)] \quad (3.1)$$

where:

$$P = Ln\left(\frac{\text{Actual hrs}}{\text{Estimated hrs}}\right)$$

$A = Ln$ (Electrical density in ft^2/worker)
$B = Ln$ (Nonelectrical density in ft^2/worker)
$Ln =$ Natural logarithm

The regression analysis is shown in Table 3.7, and the analysis of variance is shown in Table 3.8. The analysis of the residual is shown in Figure 3.8. The residuals are normal, the scatter plot did not have a pattern and the residuals were within tolerance.

One of the most relevant measures of statistical performance is the R^2 ratio. The closer the R^2 ratio is to 1.0, the more variation is explained by the model. The more variation explained by the model, the better the model performs. The model depicted in Equation 3.1 has an R^2 of 0.757. However, there is a problem in just

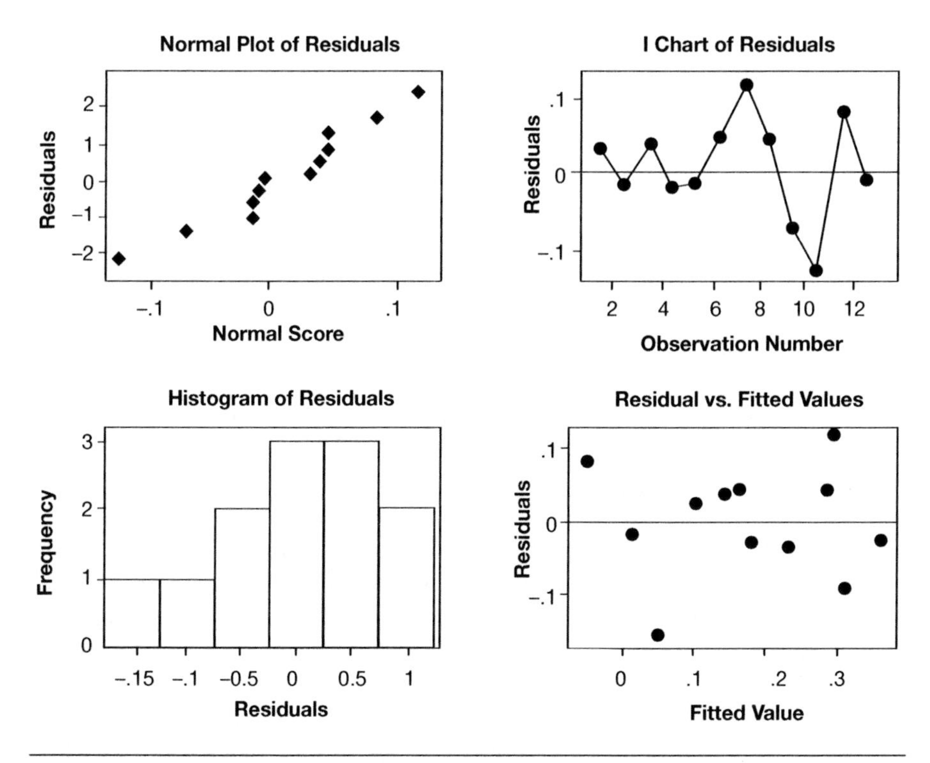

Figure 3.8 Analysis of residuals.

using the R^2 ratio as the only measure of model performance. The R^2 ratio always increases as more independent variables are added to a model. The increase in R^2 occurs even if the newly added independent variable adds no value to the model. As a result, an adjusted R^2 value is used as another statistical measure. The adjusted R^2 ratio also explains how much variation around the mean of the dependent variable is explained by the model after compensating for the number of independent variables. The adjusted R^2 only increases if the newly added independent variable helps to explain variation around the mean of the dependent variable. As with R^2, adjusted R^2 explains more total variation around the mean of the dependent variable as its value approaches 1.0. The model depicted in Equation 3.1 has an adjusted R^2 of 0.618.

Another important statistical value is the Cp test statistic. The Cp value indicates if a model needs more independent variables. If a model has all the important independent variables, on average, Cp will be equal to the number of independent variables plus 1. The model depicted in Equation 3.1 has a Cp value equal to 7.1. The Cp value was higher than the number of independent variables, 4, plus 1; however, it was significantly closer than all of the other models generated.

The overall F test measures whether the coefficients of the independent variable do not equal zero. There are two values of importance when looking at the overall F test: the F value and its P value. The higher the F value, the more likely the coefficients do not equal each other. The P value measures how much risk is being assumed by the researcher when accepting the hypotheses that the independent variables are working together to explain the dependent variable. We consider a P value of 0.10 as an acceptable amount of risk; therefore, if the P value for an independent variable is less than 0.10, the null hypothesis that the independent variables make no contribution is rejected. As seen in Table 3.7, the model depicted in Equation 3.1 generated an F value of 5.45 and a P value of 0.026. Both of these values were within acceptable tolerance.

The T test measures the performance of each individual independent variable's capacity to explain the dependent variable. The T test is measured with a T value and a P value. As with the F value, the higher the T test, the more the independent variable explains the performance of the dependent variable. The P value once again measures the amount of risk that a researcher is willing to accept. As seen in Table 3.7, all T tests for independent variables had absolute values between 1.66 and 3.89. The P values, in Table 3.7, ranged from 0.006 to 0.140. Both the P values and the T test values were within tolerance.

Residual analysis is an important step in verifying if a regression model is adhering to the assumptions of regression analysis and/or detection of outliers in the data set. As shown in Figure 3.8, there are four graphs to observe when examining residuals. The first graph is a Normality Plot of the Residuals. The residuals from a regression analysis should have a normal distribution because of the assumptions that errors should be randomly distributed around the regression line. Residuals that are normally distributed should have a straight line in the Normal Plot of Residuals. If this plot is curved or has gaps between points the residuals may not be normal or the data set may contain outliers. The Normality Plot, shown at upper left of the figure, appears to be straight and does not contain significant gaps or bends.

The second graph, the Histogram of the Residuals, at lower left, also checks for normality in the residuals. This graph should have the normal distribution's bell shape. If the graph is not bell shaped, the data may need to be transformed. Figure 3.8 shows that the Histogram of the Residuals has a bell shape.

The third graph is the I Chart of Residuals in the upper right corner of the figure. This graph helps detect residuals that are out of tolerance and displays how the residuals change over time. In this study, the I Chart of Residuals is used for

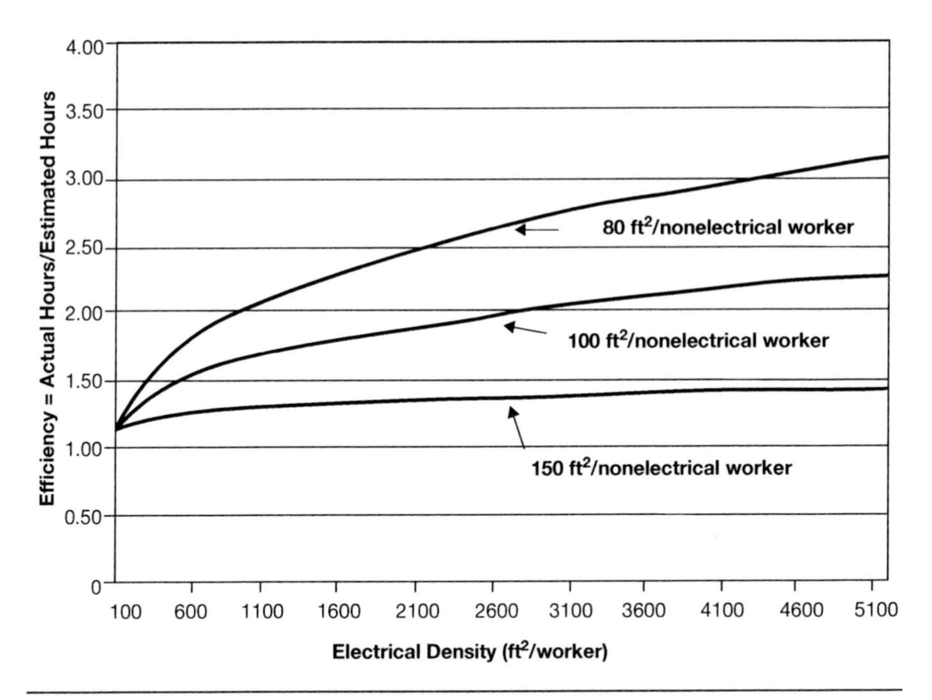

Figure 3.9 Efficiency vs. electrical density for nonelectrical density below 300.

detecting residuals that are out of tolerance. In Figure 3.8, the I Chart of Residuals shows that all residuals are within tolerance.

The final graph is a Scatter Plot of the Residuals versus Fitted Values, shown at the lower left of the figure. This graph has the model-generated values as its independent variable, horizontal axis, and the residuals as its dependent variable, vertical axis. This graph should be a scatter plot with no pattern to it. A pattern would indicate that the model has a nonconstant variance, an error in the analysis that would require data transformation, or a need for additional independent variables. The scatter plot of the Residuals versus Fitted Values also helps in the discovery of outliers. If one point or group of points are clustered far away from the majority of the data, these points should be investigated. In Figure 3.8, the Residuals versus Fitted Values plot has no detectable pattern.

Equation 3.1 is depicted graphically in Figures 3.9 and 3.10. These figures have electrical density, in squared feet per worker (ft^2/worker), on the horizontal axis; while the vertical axis has labor efficiency expressed as a ratio of actual hours to estimated hours. Both figures have a series of nonelectrical density curves expressed in squared feet per nonelectrical worker. Figure 3.9 shows nonelectrical densities below 300 ft^2/worker, and Figure 3.10 shows nonelectrical densities above 300 ft^2/worker.

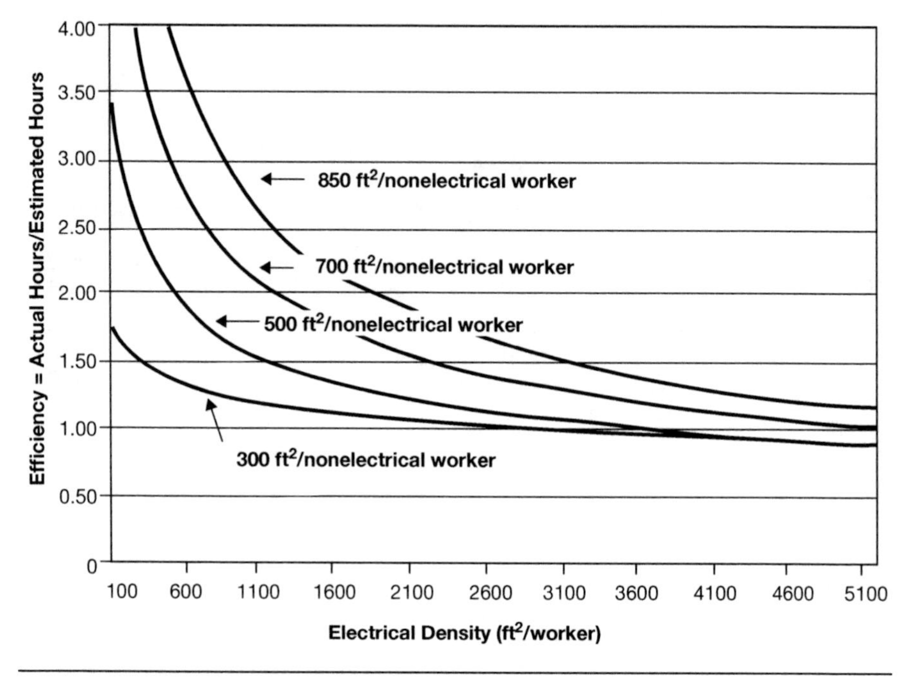

Figure 3.10 Efficiency vs. electrical density for nonelectrical density above 300.

MODEL VALIDATION

A model should be evaluated to determine its robustness. There are two primary methods used to test the performance of a model. The first test that can be performed is to validate the model internally. The second method is to compare the model to previous studies. This section presents these two primary methods of model valuation and their results.

Test Points and Cross-Validation Procedures

Test points and cross-validation procedures are used to test the predictive capacity of a model without comparing the model's performance to previous studies. After the creation of our model, new data points were requested from contractors and used to evaluate the predictive power of the model. These points are called test points. The model should predict 90% of the test points when using a 90% prediction interval. Table 3.9 contains demographics on the test points used.

The density values were entered into Equation 3.1, and the results were converted from natural log. Figure 3.11 compares the model's 90% predictive interval to the true value of the test points. The horizontal axis shows the test points used in the validation procedure. The vertical axis shows the ratio of actual hours to

Table 3.9 Test Point Demographics

Test Point	Electrical Density (ft^2/worker)	Nonelectrical Density (ft^2/worker)	True Ratio of Actual to Estimated Hours
1	574.5	208.5	1.10
2	4,087.8	645.4	1.00
3	6,808.5	513.8	1.15
4	32,026.0	412.6	0.98

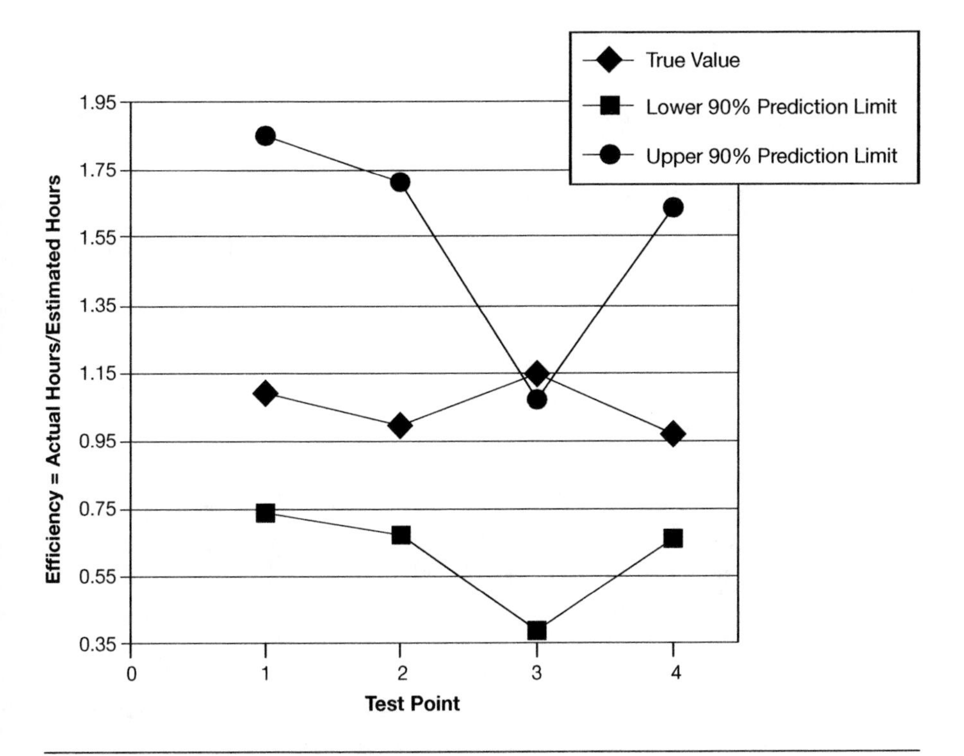

Figure 3.11 90% Predictive interval of the test points.

estimated hours. There are three lines on Figure 3.11. These lines are the upper 90% prediction limit, the test point's true ratio of actual to estimated hours, and the lower 90% prediction limit. Test point 3 was above the 90% prediction interval, but it should also be noted that test point 3 was outside the range of the data used to create Equation 3.1.

A cross-validation procedure was also used in this study. In cross-validation procedures, a single data point from the original regression model is removed and

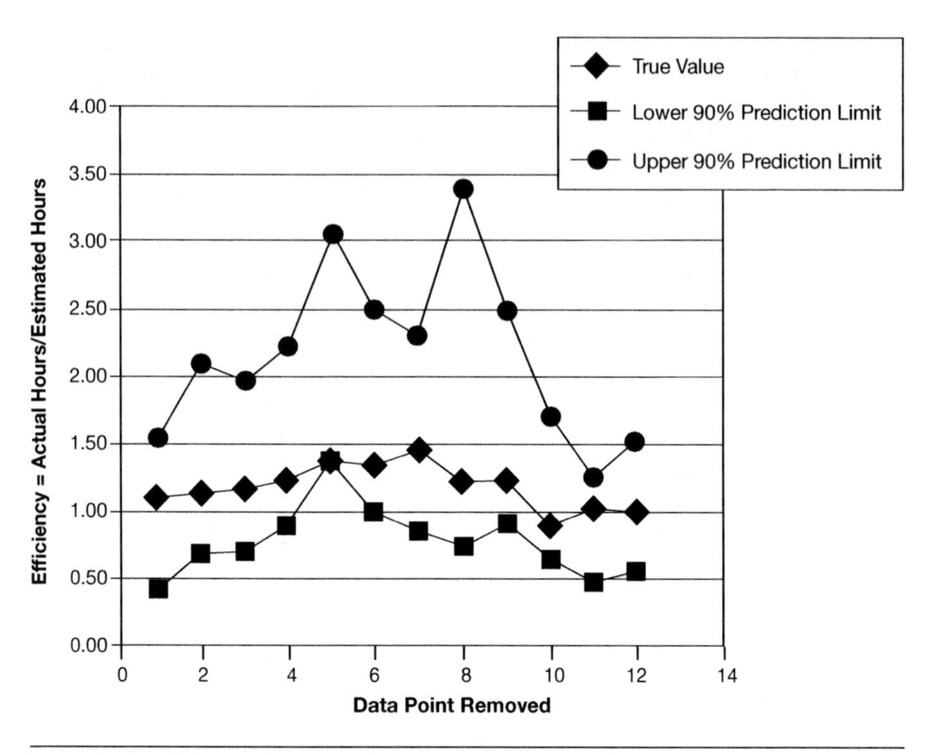

Figure 3.12 90% Predictive interval for cross-validation procedure.

a new regression is created. Then the new regression is used to predict the data point that was removed. This process of removing a data point, creating a new model and predicting its value was performed for all of the data points.

In this study, a 90% prediction interval was used to judge the performance of each of the newly generated models. If the model is performing satisfactorily, 90% of the data points should lie within the 90% prediction interval for the model. Figure 3.12 is a graphical depiction of the 90% prediction intervals for the cross-validation procedure. On the horizontal axis, the number of the data point removed during the cross-validation procedure is shown. On the vertical axis is the ratio of actual to estimated hours that is shown. The figure has three curves: the upper 90% prediction limit, the true ratio of actual to estimated, and the lower 90% prediction limit. Data point 5 falls just outside of the 90% prediction interval. In fact, Figure 3.12 shows point 5 almost on top of the lower 90% prediction limit. This point represents 6.25% of the total data points. Given the range that the model predicts, the results of the cross-validation procedures are considered satisfactory.

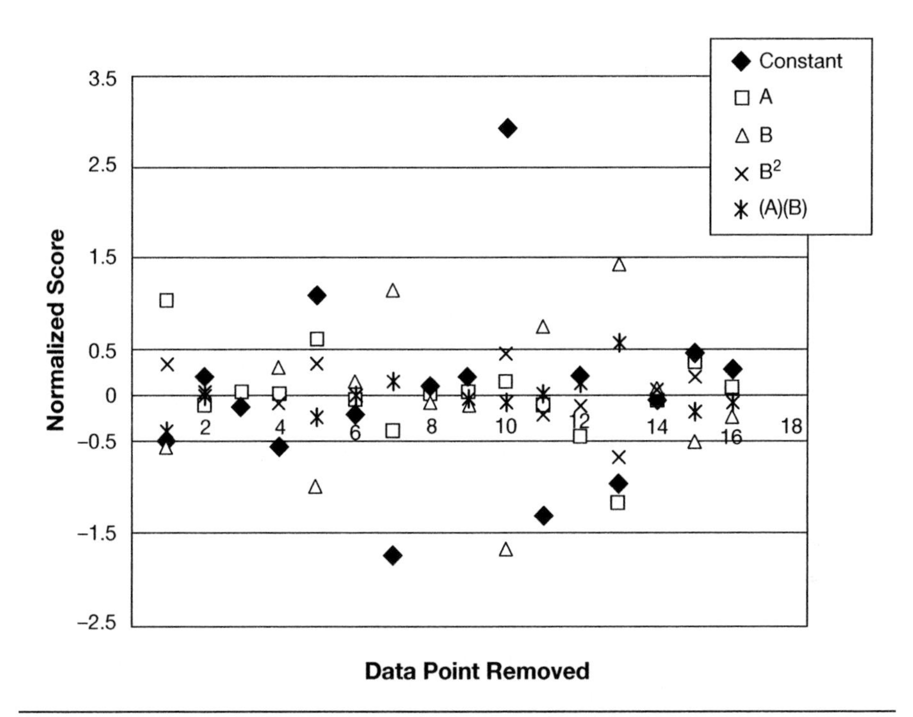

Figure 3.13 Normality plot for parameters in cross-validation procedure.

Another method for viewing the results of the cross-validation procedure is to evaluate the normalized results of the y-intercepts and coefficients for each of the regressions performed during the cross-validation procedures. It is the change in constants and coefficients that is of importance. If the data set contains many outliers, the constants and coefficients from each of the newly generated regression will be unstable. Instability of the constants and the coefficients would make it difficult to create an accurate model with the overall data. To aid in viewing, the change of the constants and coefficients data were normalized. When using normalized results, 99% of the results should be between 3 and −3.

Figure 3.13 has the normalized results of the cross-validation procedure. The horizontal axis shows the number of the data point removed. The vertical axis shows the normalized value. The data points plotted on Figure 3.13 are the constants and coefficients for each of the regressions generated during the cross-validation procedure. As shown in Figure 3.13, the constants and coefficients for all of the models remain relatively constant throughout the cross-validation procedure. The natural log of the electrical density multiplied by the natural log of non-electrical density has the largest range of all the normalized values. The range was from −1.75 to 2.83.

Table 3.10 Regression Analysis for Total Density Model

Predictor	Coefficient	Standard Deviation	T Value	P Value
Constant	-4.642	4.300	-1.05	0.314
C	2.108	1.724	1.22	0.243
C^2	-0.216	0.164	-1.31	0.211

Table 3.11 Analysis of Variance for Total Density Model

Source	D.F.	S.S.	M.S.	F Value	P Value
Regression	2	0.4762	0.2381	1.75	0.212
Error	13	1.7670	0.1359		
Total	15	2.2432			

Comparison to Previous Studies

The next primary method of validation is to compare the model to previous studies, otherwise known as external validation. There are several studies on stacking of trades or subjects similar to stacking of trades, but none of the studies presented any supporting data. Those studies used different methodology than what was used in this study. This made it difficult to compare Equation 3.1. to previous studies.

There are several studies on subjects that closely relate to stacking of trades. We mentioned several of these studies at the beginning of this chapter. The most relevant kind of such studies are site density studies. These studies use the total work area divided by the total number of workers within the area. Equation 3.1 divides labor into electrical and nonelectrical when developing a productivity model. Therefore, a total density model was developed for comparison to the previous site density studies. This model was only developed so that the collected electrical data could be plotted against the total density models. The results of the regression analysis can be viewed in Tables 3.10 and 3.11. The regression procedure is the same used to generate Equation 3.1. The resulting model is shown in Equation 3.2.

$$P = -4.642 + 2.108(C) - 0.2161(C^2) \tag{3.2}$$

where:

$$P = Ln\left(\frac{\text{Actual hrs}}{\text{Estimated hrs}}\right)$$

$C = Ln$ (Total density in ft^2/worker)
Ln = Natural logarithm

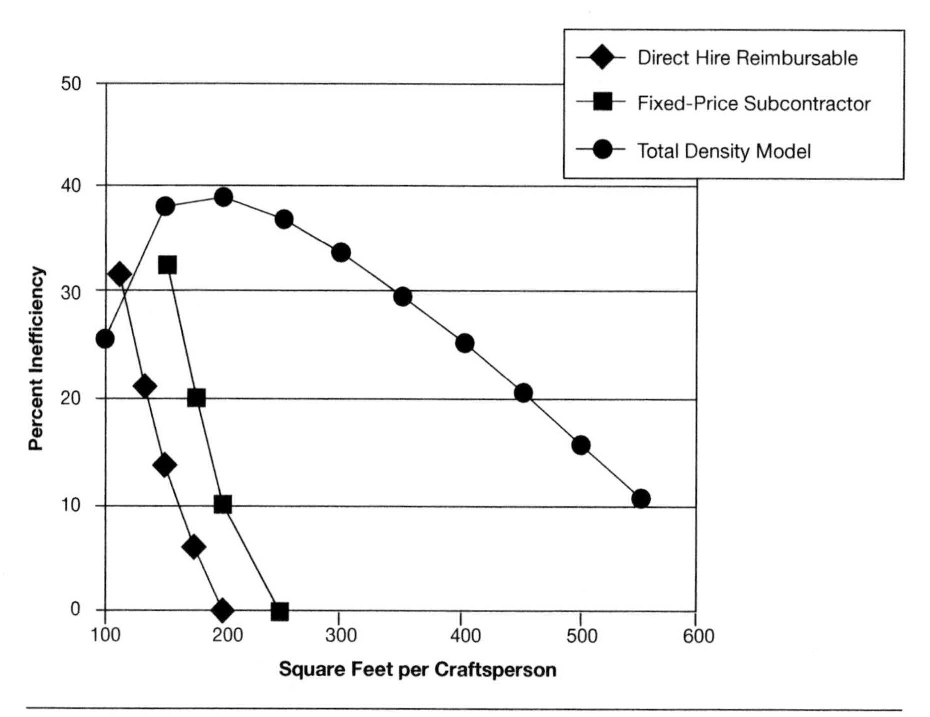

Figure 3.14 Total density model comparison to Mobil Oil study.

The model also generated the following results: $R^2 = 0.212$; R^2 adjusted 0.091; S = 0.3687; Cp = 4.0. This model failed all of the statistical tests used with Equation 3.1; however, it was the best-performing model that could be generated using total density for an independent variable.

We mentioned two site density studies at the beginning of this chapter: the Mobil Oil study and the Smith (1987) study. Figure 3.14 shows the comparison between Mobil's site density study and the total density model obtained with Equation 3.2. Figure 3.15 is a comparison between Smith's models and the site density model. The model generated by Equation 3.2 has a significantly larger range on both the horizontal and the vertical axis than these two previous studies.

Each of the figures has a different vertical axis. Figure 3.14 used percent inefficiency as its dependent variable. Percent inefficiency is the amount of productive time lost or the drop of productivity from a fixed theoretical value or estimated rate of production. Figure 3.15 directly measures productivity. Productivity is defined as actual production divided by a fixed theoretical value or estimated rate of production.

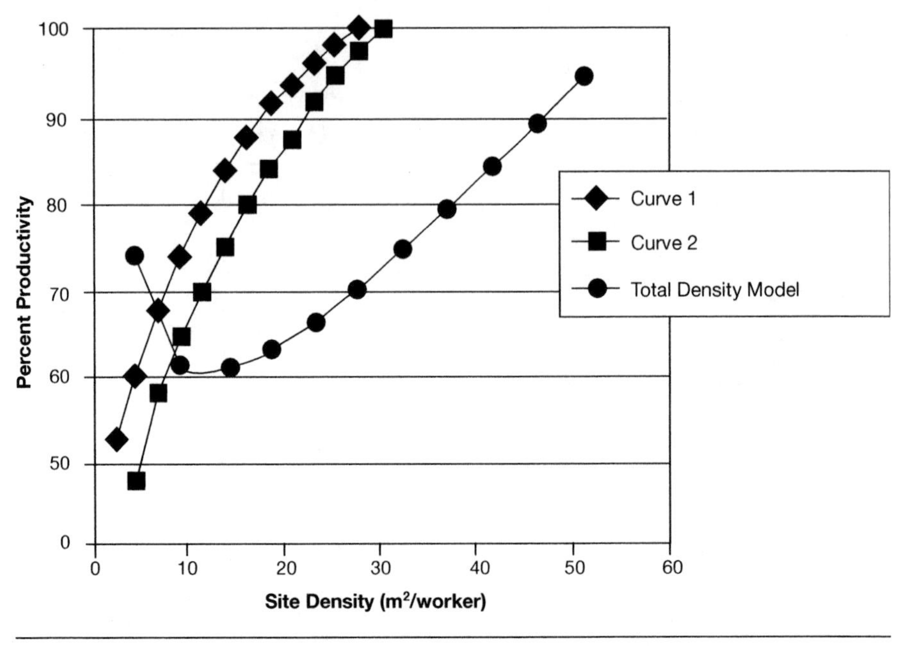

Figure 3.15 Total density model comparison to Smith study.

In each of the figures, the total density model was plotted on the graph. To plot the total density model on the graph, the results of the total density model were converted into percent inefficiency and productivity.

Figure 3.14 is a comparison between the total density model (Equation 3.2) and the Mobil site density study. It appears that the two studies do not have the same trends over the range of the Mobil study. The generated total density model crosses the horizontal axis at a higher total density value. The generated total density model also has a higher amount of productivity loss than the Mobil study.

Figure 3.15 compares the total density model to the Smith model. Once again, the total density model has a significantly different curvature than Smith's work. These models generated similar values around 10 m^2 per total workers density; however, the total density model indicates that total productivity is not achieved until after 60 m^2 per total worker density is reached at a construction site. Smith's curves indicate that full production is achieved at 30 m^2 per worker.

As explained at the beginning of this chapter, it is difficult to compare the site density studies because of the differences in the vertical and horizontal axis used. Figure 3.16 shows all the site density models uncovered during the performance of this study, the total density model developed from this study (Equation 3.2) and the 90% prediction interval of the total density model. As

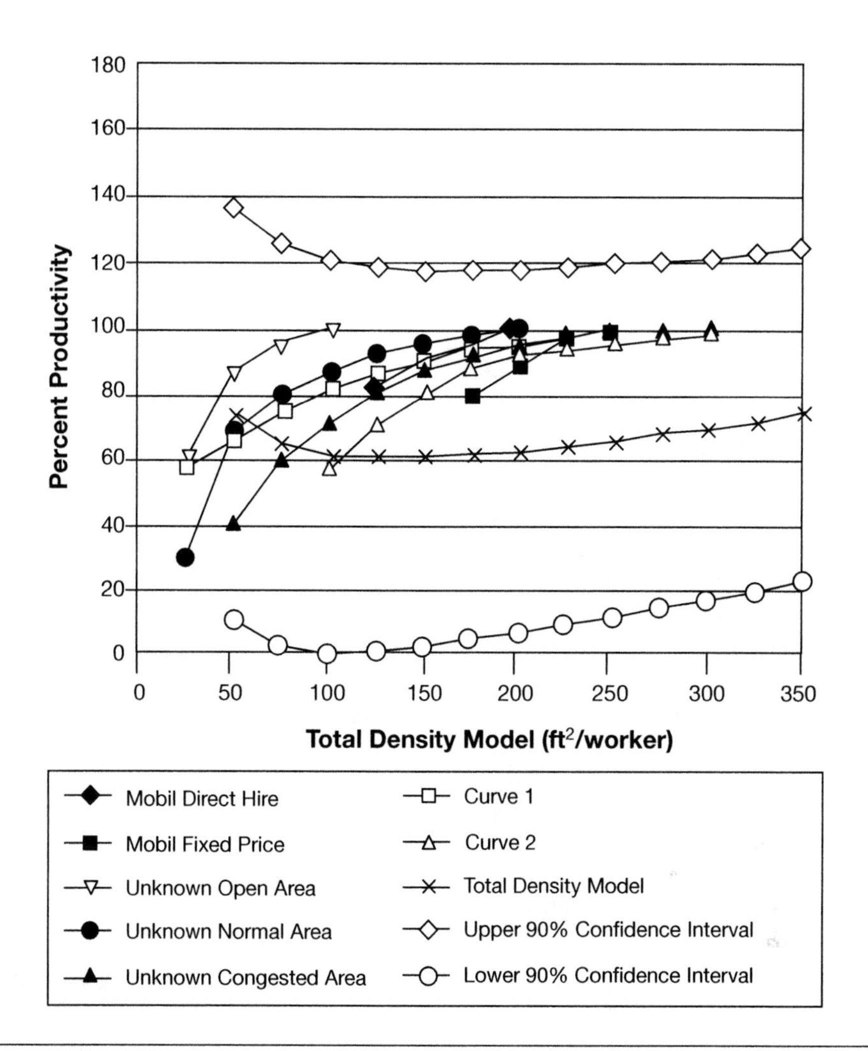

Figure 3.16 Comparison of total site density model (Equation 3.2) to previous studies.

shown in Figure 3.16, all previous studies fall within the 90% prediction interval of the total density model.

The original model developed (Equation 3.1) was also compared to the overmanning study from the Army Corps of Engineers (1979). Figure 3.17 shows the percentage of overcrowding occurring in a work area on the horizontal axis. Percent overcrowded is calculated by taking the total change in all crafts and dividing it by the previous total number of crafts. The vertical axis in Figure 3.17

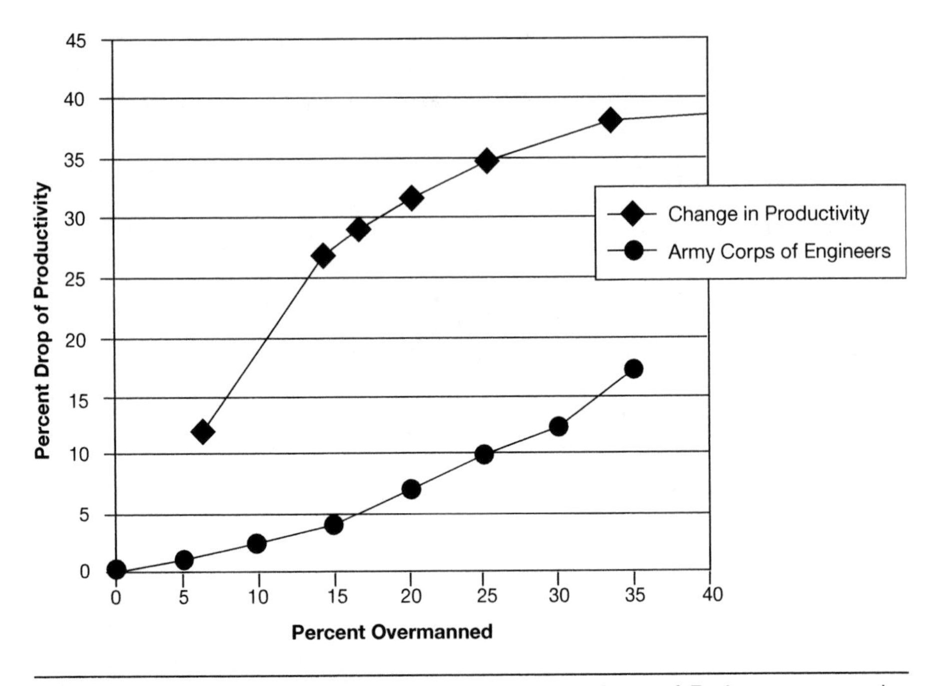

Figure 3.17 Comparison of model (Equation 3.1) to Army Corps of Engineers overmanning study.

is productivity, in percent, expressed as output of the crew divided by a theoretical maximum or estimated capacity of the crew.

To plot Equation 3.1 on Figure 3.17, electrical and nonelectrical density were set with an equal density value. Then both electrical density and nonelectrical density were equally increased to mimic overcrowding. The percentage changes in density in Equation 3.1 are plotted on the horizontal axis, while percentage change in productivity is plotted on the vertical axis. This plotting of Equation 3.1 creates a significantly higher loss of electrical productivity than that in the Army Corps of Engineers model. Equation 3.1 also has a different curvature than the Army Corps of Engineers' model.

Equation 3.1 was also compared to the information from Thomas (1994) as shown in Figure 3.18. This figure uses the percent increase of a particular trade as its independent variable (horizontal axis). This value is calculated by dividing the change in a trade's size by the previous trade size. Figure 3.18 shows the percentage drop from expected or planned productivity on the vertical axis.

Overmanning is the increase of a particular crew. Therefore, to plot Equation 3.1 on Figure 3.18, nonelectrical density would have to be constant while electrical density changes. There are many different possible values for the nonelectrical density values. Figure 3.18 has four constant values of nonelectrical densities that

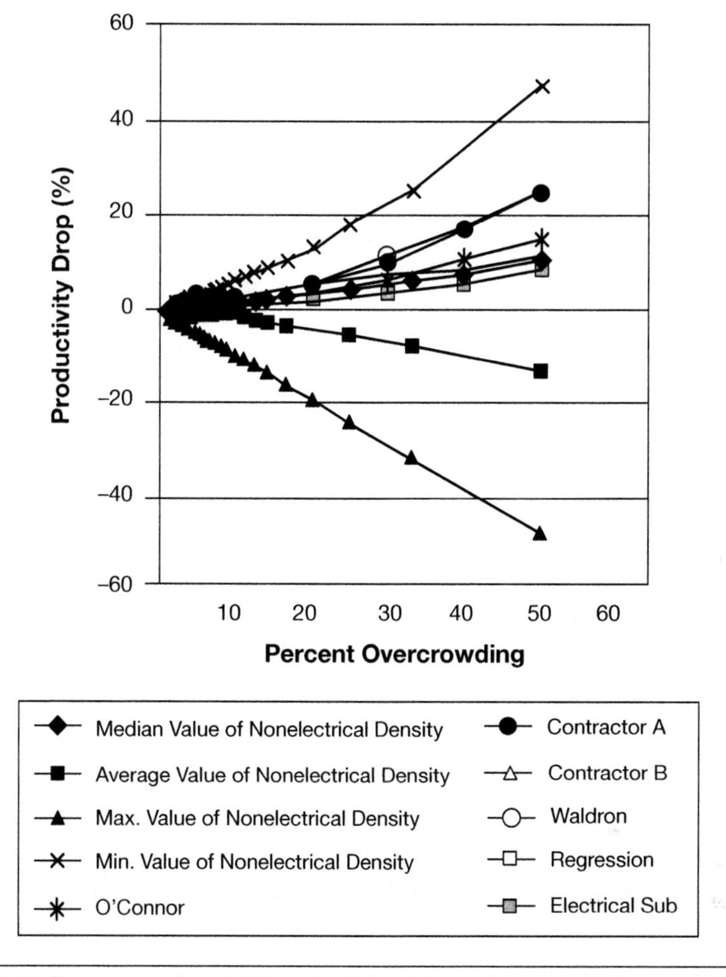

Figure 3.18 Comparison of model (Equation 3.1) to Thomas Literature Review models.

were used to plot the change of electrical crew's productivity. The constant nonelectrical densities for Equation 3.1 that were used included the minimum, the median, the average, and the maximum nonelectrical density from the quantitative data set.

The percentage change in electrical productivity was calculated by finding the difference between a new productivity level and the previous level of productivity. Then, the difference in productivity was divided by the current productivity value.

Data from Thomas (1994) and Equation 3.1 coincide over the same horizontal and vertical axis ranges. The median value of nonelectrical density, 166.3 ft^2/worker, plots directly through the Thomas review. The average value of nonelectrical density of 282.3 ft^2/worker and the maximum nonelectrical density value of 856.0 ft^2/worker indicates an increase of electrical labor productivity. The minimum value of nonelectrical density 89.1 ft^2/worker has a productivity loss greater that that derived from the Thomas (1994) data.

Analysis of Validation Effort

Equation 3.1 was validated internally and externally. Internal validation consisted of test point and cross-validation. Equation 3.1 predicted three out of four test points. The one test point that was not predicted was from data that was beyond the range of the data used to develop the model. With the cross-validation procedure, only one point fell out of the predictive capacity of Equation 3.1. The results of the internal validation were promising.

It was difficult to make an accurate and just comparison between Equation 3.1 and previous studies in areas related to stacking of trades. Previous studies did not employ a methodology similar to that of Equation 3.1. To plot against the site density studies, a new model was created based on total site density (Equation 3.2). This model poorly explained the variation in electrical productivity. The total density model plotted over a significantly different range than the previous site density studies. However, when the total site density model (Equation 3.2) was plotted against all previous site density studies uncovered during the performance of this study, it was clear that all previous studies fall within the 90% prediction interval of the total density model.

Other assumptions were also made so that the electrical productivity data could be plotted against the information from other studies related to overmanning and overcrowding. Equation 3.1 plotted significantly different productivity losses for electrical contractors than that obtained in the Army Corps of Engineers' work. However, when Equation 3.1 was compared to Thomas's review of overmanning, all overmanning studies were enveloped by the maximum and minimum values of the model (Equation 3.1).

APPLICATION OF THE MODEL

It should be noted that the application of the regression function created in this study (Equation 3.1) is limited to the maximums and minimums of electrical density, nonelectrical density, and the ratio of actual to estimated hours captured to perform this study. Four pieces of information are needed to use Equation 3.1: (1) estimate of labor hours without considering the effect of stacking of trades, (2) a measured work area, (3) number of electrical workers in the work area, and (4) number of nonelectrical workers in the work area. The following steps can be used

Table 3.12 Electrical Density Conversion

Electrical Density (ft^2/worker) (1)	Density Adjustment (2)	Equation Value 1 (3)
100	4.61	7.69
200	5.30	8.85
300	5.70	9.53
400	5.99	10.01
500	6.21	10.38
600	6.40	10.68
700	6.55	10.94
800	6.68	11.16
900	6.80	11.36
1,000	6.91	11.54
1,100	7.00	11.70
1,200	7.09	11.84
1,300	7.17	11.97
1,400	7.24	12.10
1,500	7.31	12.21
2,000	7.60	12.69
2,500	7.82	13.07
3,000	8.01	13.37
3,500	8.16	13.63
4,000	8.29	13.85
4,500	8.41	14.05
5,000	8.52	14.22
5,500	8.61	14.38

to calculate the impact that stacking of trades has on labor productivity of electrical contractors:

1. *Calculate the size of useful work area.* Useful area is the total area available minus space taken up by equipment, areas where work is not performed, and storage space.
2. *Calculate the electrical and nonelectrical density.* Electrical density is calculated by dividing the size of useful work area by the number of electrical workers. Nonelectrical density is calculated by dividing the size of useful work area by the number of nonelectrical workers.
3. *Convert electrical density.* Electrical density is converted using Table 3.12. The calculated electrical density value from Step 2 is used to

look up the density adjustment shown in column 2 and the equation value 1 shown in column 3. If the value is not shown in Table 3.12 but it is within the range of values shown, you can interpolate. However, you should not extrapolate (you should not use values greater than 5,500 for electrical density).

4. *Convert nonelectrical density.* Nonelectrical density is converted using Table 3.13. The calculated nonelectrical density value from Step 2 is used to look up the density adjustment in column 5 and equation value 2 in column 6. Once again, you can interpolate if necessary, but you should not extrapolate (you should not use values greater than 875 for nonelectrical density).
5. *Calculate the interaction term.* The interaction term, shown in Table 3.14, column 7, is calculated by multiplying the density adjustment values in columns 2 and 5. This value is then used to look up equation value 3 in column 8.
6. *Calculate the value of A.* Equation values 1, 2, and 3 from columns 3, 6 and 8 are entered into Equation 3.3:

$$A = 3.29 + \text{Equation value 1} + \text{Equation value 2} + \text{Equation value 3} \quad (3.3)$$

7. *Convert the value of A.* In Table 3.15, the value of A is shown in column 9 and is used to find the efficiency value in column 10.
8. *Calculate the work duration.* The original estimated work duration is multiplied by the efficiency value shown in column 10, Table 3.15, to arrive at a work duration that has been adjusted for stacking of trades.

Example 3.1: XYZ Electric has entered into a design-build contract for the Stable Office Complex in the City of Marathon. After discussions with the mechanical contractor and the general contractor, XYZ Electric has discovered that stacking of trades is going to occur in the mechanical/electrical room. The mechanical and the prime contractors will be performing additional work in the same room at the same time that XYZ Electric was planning to perform some conduit work. XYZ Electrical did not take this into account when estimating the amount of time, 255 work hours, needed to complete its work in the mechanical/electrical room. XYZ Electric needs to adjust its time estimate to compensate for the stacking-of-trades effect on electrical labor production so that it can generate an accurate and achievable work schedule.

1. The first step is to calculate the size of work area from the project's site plans. The electrical/mechanical room is 100 by 80 feet. This would generate a total area of 8,000 square feet. The electrical/mechanical

Table 3.13 Nonelectrical Density Conversion

Electrical Density (ft^2/worker) (1)	Density Adjustment (5)	Equation Value 2 (6)
80	4.38	-4.38
90	4.50	-4.23
100	4.61	-4.08
125	4.83	-3.75
150	5.01	-3.44
175	5.16	-3.16
200	5.30	-2.88
225	5.42	-2.62
250	5.52	-2.40
275	5.62	-2.16
300	5.70	-1.96
325	5.78	-1.76
350	5.86	-1.55
375	5.93	-1.37
400	5.99	-1.20
425	6.05	-1.03
450	6.11	-0.86
475	6.16	-0.72
500	6.21	-0.57
525	6.26	-0.42
550	6.31	-0.26
575	6.35	-0.14
600	6.40	0.02
625	6.44	0.15
650	6.48	0.27
675	6.51	0.37
700	6.55	0.51
725	6.59	0.64
750	6.62	0.74
775	6.65	0.84
800	6.68	0.95
825	6.72	1.09
850	6.75	1.19
875	6.77	1.26

Table 3.14 Interaction Term

Interaction Term (7) (7) = (2)*(5)	Equation Value 3 (8)
20	-6.45
22	-7.09
24	-7.74
26	-8.38
28	-9.03
30	-9.67
32	-10.32
34	-10.96
36	-11.61
38	-12.25
40	-12.90
42	-13.54
44	-14.19
46	-14.83
48	-15.48
50	-16.12
52	-16.77
54	-17.41
56	-18.06
58	-18.70
60	-19.35

Table 3.15 Efficiency Value Calculation

A (9)	Efficiency Value (10)
-0.10	0.90
0.10	1.11
0.20	1.22
0.30	1.35
0.40	1.49
0.50	1.65
0.60	1.82
0.70	2.01
0.80	2.23
0.90	2.46
1.00	2.72
1.10	3.00
1.20	3.32
1.30	3.67

room already has its two chillers and their control center installed. XYZ Electrical is not performing work inside or on the chillers. Therefore, the area containing chillers and their control equipment cannot be used by XYZ Electric and should be subtracted from the total area of the electrical/mechanical room. The chillers and related control equipment occupy 2,500 square feet. The total usable work area is 5,500 square feet.

2. The second step is to calculate electrical and nonelectrical worker density. To accomplish this, XYZ Electric has to collect information on how many electrical and nonelectrical workers are going to work in the mechanical/electrical room. XYZ Electric plans to have 5 electricians in the mechanical/electrical room. In discussions with the mechanical and general contractors, XYZ found out that the total number of nonelectrical workers in the mechanical/electrical room is going to be 22. Electrical density is calculated by dividing the usable work area, 5,500 ft^2, by 5, for an electrical density of 1,100 ft^2/worker. Nonelectrical density is calculated by dividing the usable work area, 5,500 ft^2 by 22, for a nonelectrical density of 250 ft^2/worker.
3. The value for electrical density (1,100 ft^2/worker) is used in Table 3.12 to find the density adjustment value in column 2 and the equation value 1 in column 3. For the given electrical density of 1,100 ft^2/worker, the density adjustment value is 7.00 and the equation value 1 is 11.70.
4. The value of nonelectrical density (250 ft^2/worker) is used in Table 3.13 to find the density adjustment value in column 5 and the equation value 2 in column 6. For the given nonelectrical density of 250, the density adjustment value is 5.52 and the equation value 2 is –2.40.
5. The two density adjustment values of 7.00 and 5.52 are multiplied together to obtain 38.64. This value of 38.64 is used in Table 3.14, column 7, to find the equation value 3 in column 8. Table 3.14 does not have a value of 38.64, but the equation value 3 can be interpolated by using the values of 38 and 40. For 38, the equation value 3 is –12.25 and for 40, the equation value 3 is –12.90. Therefore, a simple linear interpolation provides an equation value 3 of –12.46 for an interaction term of 38.64.
6. Equation values 1, 2, and 3 are placed in Equation 3.4 to calculate the value of A ($A = 3.29 + 11.70 - 2.40 - 12.46 = 0.13$).

7. The value of 0.13 must be converted using Table 3.15. Column 9 does not have a value for 0.13, but it can be linearly interpolated between 0.10 and 0.20. In this case, the interpolated value is 1.14. Therefore, the efficiency factor is 1.14.
8. The efficiency factor of 1.14 is multiplied by the estimated hours of 255 to arrive at an activity duration that has been adjusted for stacking of trades. XYZ Electric can expect to spend about 291 hours in the mechanical/electrical room, and they should plan accordingly.

CONCLUSIONS

Even though this study was developed specifically for electrical contractors, the principles of stacking of trades apply to other specialty areas as well. However, if mechanical, sheet metal, piping, or other contractors want to calculate the effects of stacking of trades on their productivity, they should replicate this study specifically for their trades so that they can develop their own versions of Equation 3.1.

WEB ADDED VALUE

Stacking.xls

This spreadsheet implements Equation 3.1 using the data from Example 3.1. You can use this spreadsheet as a template to enter your own data, since all formulas are included.

This book has free material available for download from the Web Added Value™ resource center at *www.jrosspub.com*

4

UNDERSTANDING SCHEDULE ACCELERATION AND COMPRESSION

Dr. H. Randolph Thomas, *The Pennsylvania State University*
Dr. Amr Oloufa, *The Pennsylvania State University*

INTRODUCTION

Schedule acceleration and compression is a serious problem for contractors. While schedule acceleration takes many forms and every project is unique, contractors would agree that schedule acceleration or compression will cost the contractor more money. The goal of this chapter is to provide guidance and strategies on how to minimize the economic consequences of schedule acceleration. The emphasis is on labor resources, since it is the most difficult resource to manage and quantify.

When confronted with schedule acceleration or compression, managers have three options for increasing the labor hours: overtime, overmanning, and shift work. Each has advantages and disadvantages, as illustrated in Table 4.1. There are numerous overtime schedules, and these can generally be implemented rather quickly. However, research shows that labor productivity declines with the use of an extended overtime schedule (NECA 1989). Contractors can also hire more

Table 4.1 Advantages and Disadvantages of Options for Increasing Labor Hours

Recognized Approach	Advantages	Disadvantages
Scheduled Overtime	Can be done quickly and for short periods of time.	Working prolonged periods of overtime will lead to fatigue, low morale, and possibly increased accidents.
	Perhaps the least costly of the three options because of the way the payroll burden is determined.	May need to develop an individual work rotation plan to avoid overtime fatigue.
	Owners may agree to pay the premium portion of the labor cost.	
Hiring of More Craftspeople (Overmanning)	Can avoid the overtime problems of fatigue.	It takes longer to get up to speed because the workforce will be inexperienced with the site.
		New hires may be poorly trained.
		The cost per unit work hour will be more than with overtime.
		Site congestion may become a problem.
Shift Work	Will usually alleviate site congestion problems.	Not all work is suited for a second or third shift.
	Can do work with special requirements during off hours.	Coordination between shifts is more difficult.
	May be able to minimize the cost of equipment rentals ($/day).	

craftspeople. Overmanning is often used in conjunction with scheduled overtime. This combination places added burdens on supervision to provide the resources, minimize the effects of an ever-increasing frequency of disruptions, and maintain the orderly progress of the work. Craft congestion and stacking of trades at the work face is another important concern. The cost implications of adding additional labor are generally greater than scheduled overtime. Another way to add additional labor is to initiate a second or third shift. However, while another shift may relieve congestion concerns, coordination between shifts is critical. The selection of what work to be performed on the second shift is not a trivial matter and requires careful consideration. The resource availability problem is still present.

THE EFFECTS OF SCHEDULE ACCELERATION

Numerous organizations have published the results of scheduled overtime on labor efficiency. Among these are the National Electrical Contractors Association (NECA 1989), the Mechanical Contractors Association (MCA 1968), the Business Roundtable (BRT 1980), and the Construction Industry Institute (Thomas and Raynar 1994). These and other reports are consistent in that a 50-hour per week work schedule will result in a loss of labor efficiency of about 15% after four to five continuous weeks of overtime. In the short term, one principal reason for lost of efficiency is the difficulty in providing adequate resources to the crafts—that is, materials, equipment, tools, and information. Beyond four to five weeks, fatigue becomes an ever increasing problem. After 10 to 12 weeks of a prolonged overtime schedule, losses of efficiency can easily exceed 35% to 40%.

The literature is much less definitive regarding the effects of overmanning on labor efficiency. Two of the more often-cited reasons for losing efficiency are congestion or stacking of trades and dilution of supervision. Of the studies published, the consensus seems to be that efficiency losses are in the range of 10% to 20% (Thomas and Raynar 1994). However, little is known about the sources of data or how the studies were done. The presence of many additional factors could mean that the actual efficiency losses are greater than were reported in the literature.

Strategies to minimize the negative effects of schedule acceleration must begin with an understanding of the environment in which the work is to be done. Research has identified the following major characteristics of the job environment when the schedule is accelerated (not all characteristics will be present on each job):

- Specialty contractors have limited control over the schedule; others influence what happens.
- Schedule acceleration leads to changes in the daily work plan, making it difficult to plan much beyond the next day; changes occur daily.
- Specialty contractors primarily react to the situation to make things happen.
- All the schedule float is gone.
- There are many more intermediate deadlines.
- Changes often cause or accompany acceleration; the demands on the information and material supply network increase.
- Foremen, who normally do the planning on a weekly basis, are stretched to the limit such that the short-range planning function is greatly hampered.
- The turnaround time for taking positive action is relatively short.

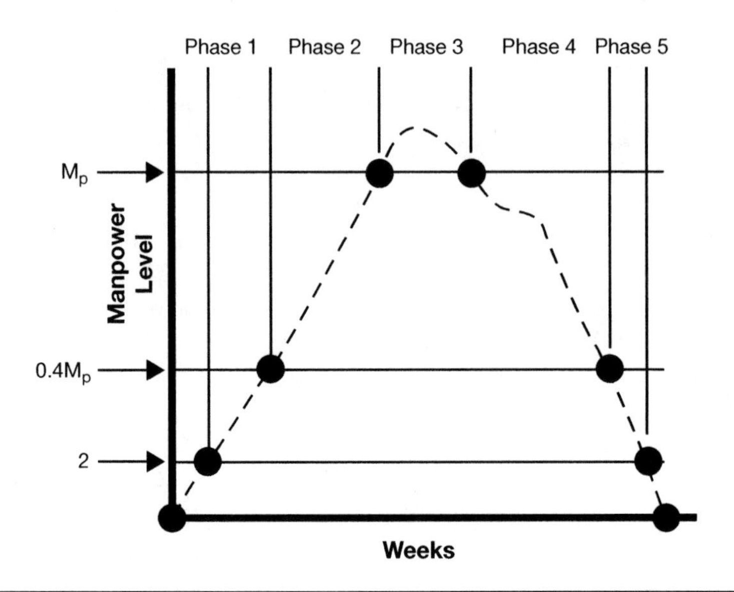

Figure 4.1 Relationship between actual manpower milestones and project phases.

An important research finding is that accelerated projects evolve through phases, and the patterns of labor inefficiency are different for each phase. Figure 4.1 shows how the manpower-level milestones are used to define the phases for indoor electrical work. The phases are a function of the number of craftspeople. To define project phases, it is only necessary for the contractor to know the planned maximum number of craftspeople, Mp. A threshold value of 0.4 Mp defines when the number of craftspeople begins to increase at an appreciable rate. The value of 0.4 Mp is rounded to the nearest integer number, but is not less than 2. For a large size workforce, this threshold value should likely be no greater than 10. At the end of the project, this threshold value also defines when the major portions of the work are completed, and the remaining work consists largely of startup and testing activities.

Phase 1 of the work begins when at least two craftspeople are assigned to the project continuously and extends until the workforce consistently reaches 0.4 Mp. During this phase, labor inefficiencies are not always readily apparent. This is due in part to the expectation that some inefficiency is inherent with starting a new project. Also, there are relatively few work hours expended in Phase 1. However, when the work is delayed and leads to the prolonging of Phase 1, unnecessary inefficiencies will occur. Following are some of the more common causes of labor inefficiencies during this phase:

- Piecemeal or out-of-sequence work means that there are frequent crew relocations from one work area to another. Piecemeal work can result from late design, limited access to work areas, delays by other contractors, environmental problems, late equipment deliveries, and many other reasons.
- Drawings approved for construction may be incomplete. Reading drawings with errors or omissions is inefficient.
- Waiting for instructions can lead to excessive idle or waiting time.
- Craftspeople must become familiar with the layout of the site, management preferences, procedure and work rules, characteristics of the drawings, and perform material shakeout.

Phase 2 begins when the number of craftspeople consistently exceeds 0.4 Mp. This phase continues until the actual number exceeds the planned maximum Mp. During this phase, there is a quickening pace as more work becomes available. When the work is behind schedule, there is an intense effort in Phase 2 to recover time. Unfortunately, in many cases, there is a continued lack of available work. There are frequent schedule changes, and drawings approved for construction are incomplete. The situation leads to the following causes of inefficiency:

- There is a disproportionate amount of piecemeal work compared to production work.
- Out-of-sequence work is common due to a significant push to make work available.
- The design drawings contain numerous errors, omissions, and ambiguities.

Phase 3 begins when the actual number of craftspeople exceeds the maximum planned number, Mp, and ceases when the actual number again reaches Mp. Because the labor force is overmanned, almost all activities feel the impact of the acceleration. The causes of labor inefficiency are somewhat unique to Phase 3. Following are some of the more common causes:

- Congestion.
- Scheduled overtime.
- Stacking of trades.
- The influx of design changes and clarification begins, if not started in earlier stages; thus, rework can be a problem.
- Dilution of supervision.

Phase 4 begins when the actual number of craftspeople again reaches the maximum planned number, Mp. Phase 4 continues until the size of the workforce consistently falls below the 0.4 Mp. In general, this terminating point coincides with 85% to 95% of the total elapsed duration. The principal reasons for losses of labor efficiency during Phase 4 are as follows:

- Stacking of trades
- Progressively less production work and more piecemeal work
- More changes and rework

Collectively, these factors make Phase 4 probably the least productive period of work. Finally, Phase 5 represents the last stage of the work where systems are being tested, continuity checks are being made, and the contractor is preparing to leave the site. This phase represents the last 5% to 15% of the duration of the work. There is very limited production work during this phase.

An analysis of numerous accelerated projects indicates that inefficiencies can occur in all phases. Claim procedures that ignore one or more phases would likely understate the total labor inefficiencies. Not surprisingly, the inefficiencies increase as the project continues. Generally, Phases 1 and 2 are the most productive and Phase 4 is the least productive.

A number of projects have been observed to oscillate between acceleration and deceleration in the labor resources assigned to the project. The deceleration periods were found to be equally unproductive as the periods of acceleration. Because of the fluctuations in workloads, the labor resources charged to the project also fluctuate. Reduced labor resources meant that sometimes projects reverted back to an earlier phase. Thus, a project could evolve from Phase 1 to Phase 2 and then back to Phase 1.

REDUCING THE ECONOMIC IMPACT OF ACCELERATION

There are several ways of reducing the economic impact of acceleration on the contractor. This section explores four approaches contractors can use to minimize the negative effects of acceleration: identification of constraints, inclusion of all costs in acceleration claims, proper documentation of facts, and the application of nontraditional management methods.

Identification of Constraints

One proactive approach prior to acceleration is to work toward removing the constraints that inhibit labor performance. Table 4.2 provides a brief synopsis of some of the possible constraints inhibiting progress, but there can be many more constraints. Unless constraints are removed, performance will surely deteriorate, and the schedule acceleration will become very costly. Identifying constraints may be best done in conjunction with the owner and general contractor.

Table 4.2 Some Possible Constraints Inhibiting Progress

Category	Possible Constraints
Changes	Since changes are an important source of schedule difficulties, the process itself is likely to be a constraint. Other possible constraints include the completeness and accuracy of the directives and delays in the approval process. Lack of a submittal priority schedule may also be a constraint.
Planning and Scheduling	The construction schedule may contain flaws and be in need of updating. Making new areas of work available may alleviate unfavorable working conditions; rethinking the sequence may improve progress.
Material Deliveries, Storage, and Cleanup	Late material deliveries may be alleviated by negotiating the relaxing of restrictive material specifications. Improvements in on-site deliveries and cleanup can also improve labor performance.
Owner-Procured Equipment	The schedule for owner-procured equipment may also need to be updated.
Information	An important constraint may be that engineering and design information cannot be obtained in an efficient manner.

Inclusion of All Costs in Acceleration Claims

The major elements of additional costs due to acceleration are labor, materials, installed equipment, construction equipment and tools, and site and home office overhead.

Labor Costs

The direct cost of labor has the following components: unit cost, unit rates, and quantities. The most obvious way for the unit cost ($/hr) to increase is for the project schedule to unexpectedly extend into a new labor agreement period where there is a wage increase. Other unplanned ways where unit costs can increase include premium time pay for overtime, shift differential pay for second and third shifts, and per diem. Another, perhaps overlooked, way the unit crew cost can increase is for the planned journeyman-to-apprentice ratio to increase.

The unit rate (hours/unit) or labor productivity can worsen because of various factors impacting labor efficiency. Some of the major root causes are as follows:

- Lack of materials, tools, and construction equipment
- Out-of-sequence work
- Lack of or erroneous information
- Congestion
- Changes and rework

- Dilution of supervision
- Working within an operating environment
- Unfavorable weather

Whenever schedules change, there is likely to be a disruption of the orderly procurement, delivery, and installation of materials and equipment. Such disruptions interrupt the normal planning process and can lead to out-of-sequence work. When this happens, foremen do not have adequate time to develop accurate bills of materials or check the completeness of materials and information for the planned work package. When schedule disruptions are acute, certain needed materials may not have been procured. This problem is compounded when the quality of the construction documents is lacking, thus resulting in more out-of-sequence work. This scenario is a common one when there are many changes to the work.

When workers have to return to a work location to complete work that could have been done earlier, they often find that the character of the work location is different than what it was when the work was originally scheduled to have been done; thus, the environment has changed. The work area may have become congested with installed components or other craftspeople, or that part of the facility may now be in operational testing or in operation. The work will now be much more difficult, resulting in more work hours than budgeted to do the same amount of work. When work is done outdoors, the weather may be a factor. If specification requirements are changed during the course of the project, more restrictive specifications and inspection criteria will impair production.

The use of construction equipment and tools may also lead to increased hours to do the same amount of work. Access to lifting equipment may be limited, thus resulting in more waiting time by the crew. In some cases, smaller, less efficient lifting equipment may need to be used. Where there is downtime for equipment or tool maintenance, labor efficiency will suffer.

Accelerated or compressed schedules can be particularly detrimental to labor efficiency because they make these kinds of problems more acute. Additionally, overtime schedules that last for more than four to five consecutive weeks will result in worker fatigue. Fatigue leads to reduced efficiency, worsening morale, increased absenteeism, and in some cases, higher accident incident rates. Increasing the size of the workforce can cause a more congested work environment. Additionally, new hires may not be as skilled as the current workforce. At a minimum, time will be needed for these new workers to become acclimated to the jobsite. Shift work poses unique problems. Unless the work for the second or third shift is carefully selected, efficiency will be reduced. These shifts are not copies of the first shift.

An important factor related to increasing or decreasing quantities is the learning curve effect. If quantities are increased, it matters if the work can be integrated within the other work, or if the crew must return to another location to install a small quantity. A decrease in quantities can also lead to increases in unit rates. If minor reductions in quantities are made, there will likely be minimal or no increase in the unit rates (loss of efficiency). However, if significant quantity reductions are made, the unit rate can increase appreciably.

Material and Installed Equipment Costs

The cost of materials and installed equipment is composed of the base cost plus transportation, insurance, and storage surcharges. The base cost of materials can increase in several ways. If more materials are purchased and are not included in the original purchase order, vendors may charge a higher unit price because they may have to initiate a new production run for the new, but smaller, requisition. If construction schedule changes dictate that the fabricator/vendor change the production sequence, higher base costs may be incurred. Lastly, if material quantities are deleted, the fabricator/vendor will need to distribute its fixed overhead cost over fewer quantities, thus necessitating a higher unit cost. If the progress in the project is slowed and material requisitions are delayed, there may also be price escalations.

If smaller or more frequent material deliveries are required, this line item will be more expensive. Insurance costs will also increase. This situation is likely to occur if schedule changes are caused by schedule acceleration. Delivery schedules may also change if the construction schedule necessitates unplanned or unanticipated alterations in the material storage areas.

Contractors many find that because of schedule acceleration, storage areas are moved or reduced or more materials are needed at the site. Thus, strains on existing storage areas may dictate that off-site storage of materials may be required. It may be necessary to rent storage lockers or space in a bonded warehouse. Storage trailers or other approaches may be required to prevent theft, and these charges should be recoverable. It is also possible that there will be more waste or spoilage. In addition, one should not overlook the cost of storing materials in the contractor's own storage yard.

Construction Equipment and Tools

Construction equipment and tool charges consist of use charges and a need for more tools and equipment to supply a larger size workforce. The hours of use may also be a factor of increased cost.

The principal reasons for increases in the unit use cost are increased maintenance cost caused by more intense use, and increased unit cost when changes require larger or different equipment and tools. On larger projects where on-site maintenance is done, it may be necessary to maintain a larger inventory of spare parts.

More craftspeople mean more equipment and tools. The notion of a bid unit cost for tools ($/unit of work) no longer applies. Equipment and tools are allocated per crew. If there are more crews, there will be more equipment and tools needs, even though the quantities may not change. On accelerated projects, delivery of materials to the work location may be an acute problem, and the contractor may need to rent more lifting equipment. A benefit may result if the same lifting equipment can be use in multiple shifts. Here the equipment may be used for twice as many hours for the same daily rental charge.

Overhead

Site overhead can include many elements of cost and is handled uniquely by each contractor. Site supervision is a key cost element. Loss of efficiency can lead to increased site supervision in several important ways. The first relates to time. If the workforce is less efficient, it will take more time to do the same amount of work. More time means more supervisory time. Second, as the schedule slips, there is a progression to a larger workforce than originally planned. A larger workforce will require more supervision, particularly at the foreman and general foreman level.

Each construction activity requires supervision and site support for material requisitioning, maintenance, scaffolding, and a variety of other tasks. The fixed costs are distributed over the contract quantities and added to the unit costs. When quantities are reduced, the contractor may have reduced labor hours required because less work is required. However, because the fixed cost is distributed over fewer quantities, the unit cost will be greater.

Home office overhead consists of two components. The first is a variable cost covering primarily the persons assigned to administer and oversee the project. As the project is accelerated, more staff and professional persons may be assigned to the project. Even though the overall project duration may be shortened, variable costs may be more than planned.

Fixed costs include nonprofessionals like clerks, secretaries, accountants, and so on. In addition, there is the cost of maintaining the home office. This category includes utilities, copy charges, insurance, and so on. These costs are usually prorated in some manner. One common way is according to the percentage of revenue generated by the accelerated project. This percentage may be lower than the actual costs.

Proper Documentation of Facts

It is widely recognized that the documentation practices on many construction projects are inadequate, untimely, or imprecise; that is, documentation frequently lacks critical dates and factual information. In particular, written communications are often not maintained properly, perhaps because of the urgency of the decision-making process. A party will have a difficult time prevailing in a dispute if that party is unable to substantiate the facts. Careful attention to documentation will permit speedy retrieval of vital information should a dispute arise. There is little doubt that many contractors fail to recover additional costs because of inadequate documentation.

An invoice showing additional charges is not sufficient; there must be a showing of entitlement—that is, a cause-effect relationship showing that the increased costs were attributable to some action or event from which the written contract grants relief. What are often missing are the facts. Exact dates, meeting participants, the content of conversations and directives, and a variety of other events may be forgotten.

Prior to construction, the contractor must review contract clauses and addenda, especially procedures for notice, approval of substitutions, disputed work, changes, warranty clauses, and so on. The specific procedures should be clearly documented. Many instances have occurred where contractors failed to recover legitimate costs simply because the procedures clearly stated in the contract were not followed.

Daily crew quantity reports are invaluable for substantiating losses of efficiency. These reports need not be complicated and should be designed so they can be completed by a crew foreman in no more than a few minutes. A report should be required from each crew. The report should note significant problems or interferences affecting the work of the crew. Furthermore, it is very important that there be a record of activity prior to the events causing the inefficiencies for which the contractor is seeking recovery. Such information can be used to establish a baseline or measured mile.

Another valuable source of information is transmittal and submittal logs. It is useful to maintain logs of submittals, requests for information (RFIs), change requests, and so on. Such logs should include dates so that with little effort, contractors can substantiate the time elapsed waiting for certain actions and how this turnaround has changed over time.

Conduct regular walkthroughs (once or twice a week), with an audio recorder, and record disruptions, stop-and-start work, and incomplete predecessor work, then transcribe this information in narrative form. Although the recorded facts may be somewhat random, the narrative creates a descriptive record of job interferences.

Photographs and videos are extremely useful in explaining problems and their impacts on a project, especially where the presentation is made to someone not familiar with the job (e.g., a mediator or arbitrator). Make certain that the photographic equipment records the time and date of the photo. Photos should record progress and job status, document problems, and show interferences. Be liberal with taking photographs.

Develop form letters with a first and last paragraph and blank space in between. Whether the information on delays or interferences is handwritten or typed, forms make it easier for the site superintendent to satisfy the contract written notice requirements. Every contract has written notice requirements, and some have fairly burdensome demands. Make a checklist for the timing and amounts of detail needed to preserve rights to equitable adjustments for delays, extra work, changes, suspensions, and so forth, and post it in the field trailer.

Application of Nontraditional Management Methods

In many schedule acceleration situations, traditional methods of project management need to be altered in order to be effective. Some methods simply will not work when the schedule is severely impacted. Thus, there is a need to take some nontraditional measures to increase the likelihood that the project will be completed with minimal economic impact. Some proven measures that can be taken are summarized on Table 4.3.

CALCULATING INEFFICIENT WORK HOURS

The methodology described in this section is based on a careful examination of labor consumption rates and labor performance of almost 400 weeks of electrical construction from five projects. The methodology involves a comparison of actual labor consumption rates to planned rates on normal or nonimpacted projects (Thomas and Oloufa 1996). The following conditions are assumed to be valid:

- Contractors adjust the level of manpower and the work schedule to be consistent with the amount of work available to be performed.
- Higher labor consumption rates resulting from overmanning and scheduled overtime mean more work is available to assign. Declining labor consumption rates mean there is a declining workload.

As described earlier, projects evolve through phases. The causes and patterns of labor inefficiency vary according to the phase, work conditions, schedule status, and amount of work available to assign. It is necessary to understand how these parameters relate to the job history before inefficient work hours can be

Table 4.3 Some Proven Measures to Mitigate Effects of Schedule Acceleration

Category	Measures
General	Monitor progress daily or weekly instead or biweekly or monthly.
	Improve communication between crews.
	Improve communications with other contractors, construction manager, and owner.
	Monitor absenteeism for detrimental trends.
	Do more planning using short-interval scheduling.
Supervision	Add more foremen and supervisors as the workforce increases.
	Stress each crew performing 100% of its work.
	Shift to smaller crews or work teams within a crew.
	Insist on greater backlogs before beginning work.
	Be selective of the work to assign to a second shift.
	Use special shift arrangements.
Material and Installed Equipment	Deliver as much material to the jobsite as practical.
	Reschedule long-lead items.
	Reassess material handling and storage areas.
	Organize a material delivery crew.
	Organize a material cleanup crew.
	Organize a setup/preparation crew.
	Use modular or preassembled components.
Tools and Construction Equipment	Increase the inventory of spare parts, hand tools, and expendables.
Information	Add supervision to identify design errors and omissions.
	Organize a special crew to handle changes.

calculated. The procedure that follows accounts for overmanning, undermanning, and scheduled overtime, since these are the only strategies that cause the labor consumption rate to change. To calculate inefficient work hours, the following information is necessary:

- The weekly labor summary indicating the actual weekly work hours and the number of craftspeople (nonsupervisory) assigned to the job
- The planned maximum number of craftspeople (nonsupervisory), Mp

Quantification Steps

The calculation of inefficient work hours is done on a weekly basis. The steps involved in performing the labor inefficiency calculations are as follows:

1. Calculate the planned normal or nonimpacted weekly labor consumption percentages using the as-planned schedule.

2. Calculate the actual weekly labor consumption percentages.
3. For each week, subtract the planned normal or nonimpacted labor consumption percentage from the actual labor consumption percentage. This difference is the weekly labor rate deviation.
4. Define the phases of work based on the actual manpower level.
5. Using the curve appropriate for each phase and value of the weekly labor rate deviation, determine the weekly performance ratio (PR) value.
6. Using the weekly PR value and actual work hours, calculate the gross weekly inefficient work hours.
7. Multiply the gross weekly inefficient work hours by the appropriate Efficiency Loss Multiplier to determine the net weekly inefficient work hours to specific events that occurred on the project.

In order to illustrate this quantification procedure, let's look at an actual electrical construction project that was accelerated.

Example 4.1: The project was a $7.5 million institutional building on a college campus. The building consisted of two wings and included an auditorium, laboratories, and classrooms. The work involved conventional electrical construction. The work interfaced with mechanical, ductwork, sprinkler, and casework contractors. The actual duration of the total project was 75 weeks, the electrical work lasted 66 of those weeks, and the actual electrician craft work hours were 18,781. The planned maximum number of electricians, Mp, was 10.

Step 1: Calculate the Planned Weekly Labor Consumption Rate

A planned schedule that is man-loaded can be used to develop the normal labor consumption rate. The normal curve is shown in Figure 4.2 using the total project duration of 75 weeks on the horizontal axis. The vertical axis is the normal labor consumption rate expressed as a percentage. The normal curve begins on week 10 to coincide with the beginning of the electrical work. The beginning and end points of the curve should correspond with the times when at least two electricians are assigned continuously to the project.

Step 2: Calculate the Actual Weekly Labor Consumption Rate

The actual weekly labor consumption rate is calculated from the weekly labor summary or manning report using the following equation:

$$\text{Weekly labor consumption rate} = \left(\frac{\text{Weekly craft work hours}}{\text{Total craft work hours}}\right)(100) \quad (4.1)$$

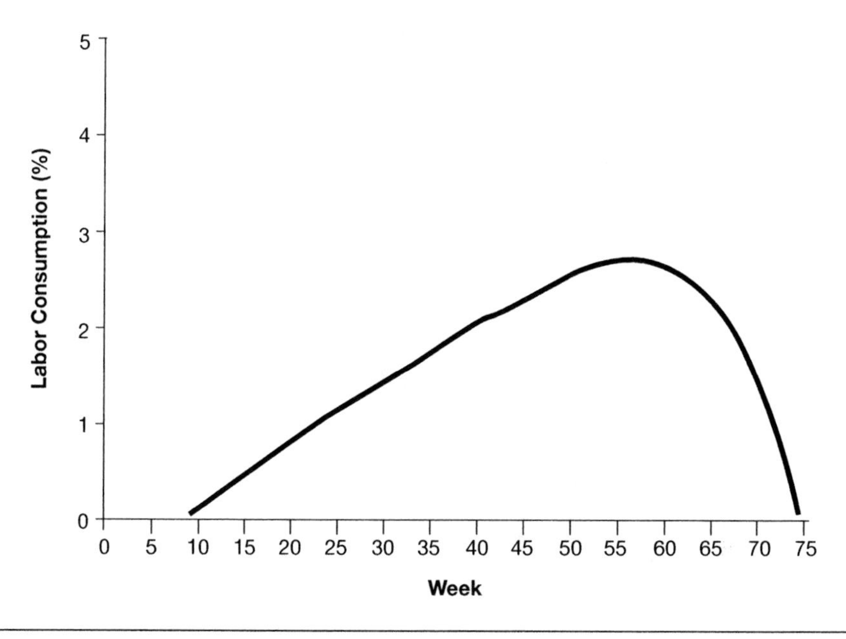

Figure 4.2 Normal labor consumption rate for example project.

The values used in Equation 4.1 are the hours occurring between weeks 10 and 75, as these are the weeks when there are two or more electricians continuously assigned to the job.

Step 3: Calculate the Labor Rate Deviation

To calculate the labor rate deviation, the planned curve is superimposed over the actual labor consumption rate curve. This is illustrated in Figure 4.3. It is important that this figure tells the story of what happens on the project. As can be seen from Figure 4.3, the work progressed normally for a brief period of six weeks; thereafter, the pace of work slowed considerably until week 39. This was followed by a period of three to four weeks where work that had been on hold was released. This period was followed by a stop-start pattern where work was put on hold and later released. At about week 65 there was a period of intense acceleration, which continued until the end of the project.

The labor rate deviation is calculated as follows:

$$\text{Labor rate deviation} = \text{Actual consumption rate} - \text{Planned consumption rate} \quad (4.2)$$

The labor rate deviation is the arithmetic difference between the two curves in Figure 4.3. Table 4.4 illustrates this calculation for selected weeks of the example

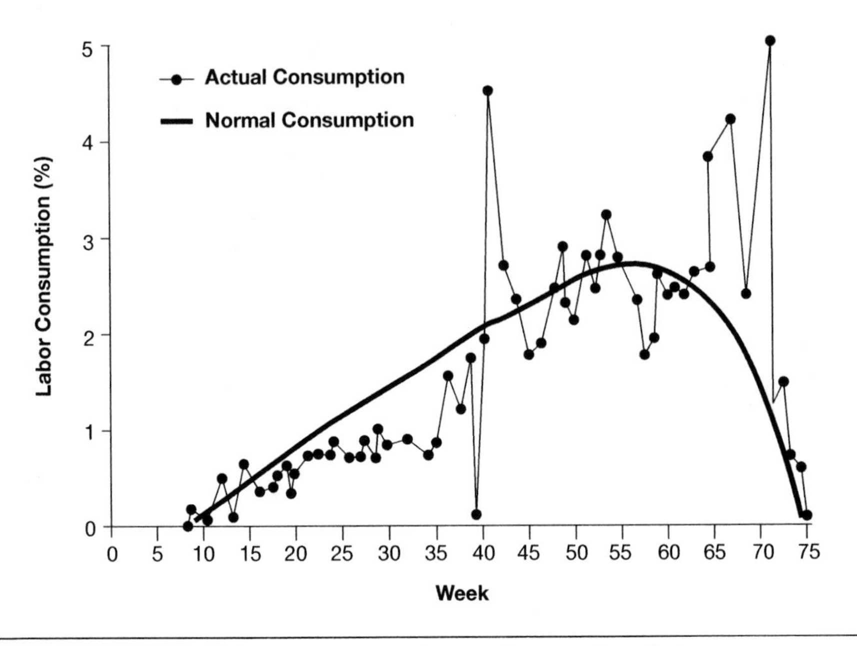

Figure 4.3 Actual and normal labor consumption rates for example project.

project. Notice that during this time frame, the pace of work as expressed by the labor rate deviation lagged behind for all but two weeks.

Step 4: Define Project Phases

Projects evolve through phases, and the quantification methodology analyzes four of these phases. Therefore, it is necessary to define the phases that are unique to each project. To define phases, it is necessary to plot the actual manpower level throughout the course of the project. It is also necessary to know the planned maximum number of craftspeople, Mp, that were to be assigned to the project.

First, we plot the actual number of electricians actually assigned to the project. The horizontal axis should be the weeks of the project. This is illustrated in Figure 4.4. Next, we determine the planned maximum number of electricians, Mp. Then we calculate 0.4 Mp. This manpower level defines the point in time when the work generally quickens. At the end of the project, this number defines the point in time when the work generally approaches completion. In calculating Mp, the number should be rounded to the nearest integer number but should never be less than 2. Two craftspeople defines the minimum number for which the quantification methodology is valid. Horizontal lines are then plotted on top of Figure 4.4 representing Mp, 0.4 Mp, and two electricians. This is shown in Figure 4.5. Where these lines intersect the actual manpower-level curve, the limits of each

Table 4.4 Calculation of Labor Rate Deviation

Week	Actual Percentage	Normal (Nonimpacted) Percentage	Labor Rate Deviation
45	1.80	2.24	-0.44
46	1.90	2.28	-0.38
47	2.33	2.31	0.02
48	2.71	2.34	0.37
49	2.20	2.37	-0.17
50	2.11	2.39	-0.28

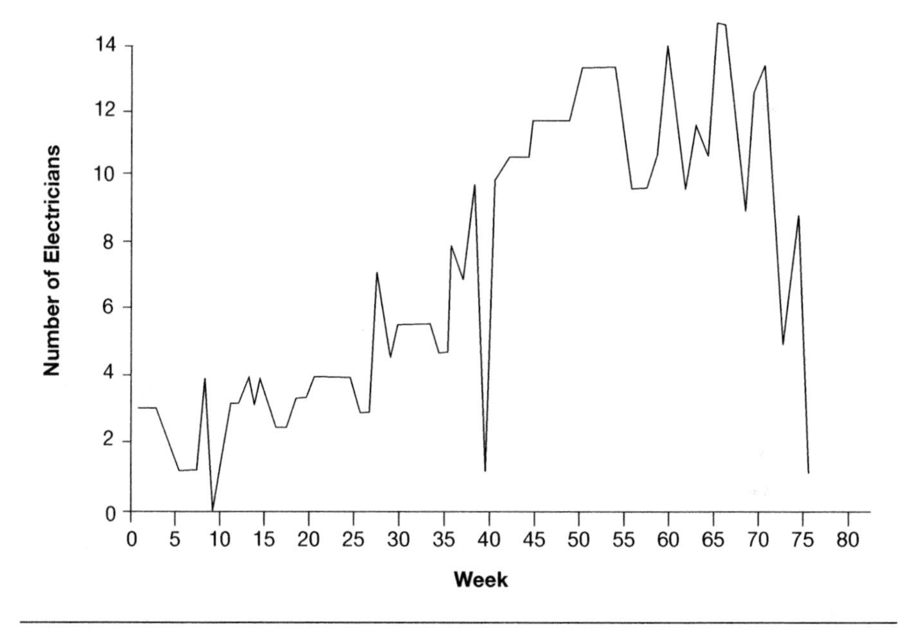

Figure 4.4 Actual manpower for example project.

phase are defined. Times when there are fewer than 2 electricians are ignored. The last phase, Phase 5, is not analyzed because there is limited production work performed during this phase.

Figure 4.5 shows how the phases are defined for the example project. The planned maximum number of electricians was 10, and 0.4 Mp is calculated to be 4. Notice that it was not until week 10 that there were two electricians assigned continuously to the project, so the first nine weeks were ignored. Where the manpower level oscillated appreciably, the project may revert back to an earlier phase.

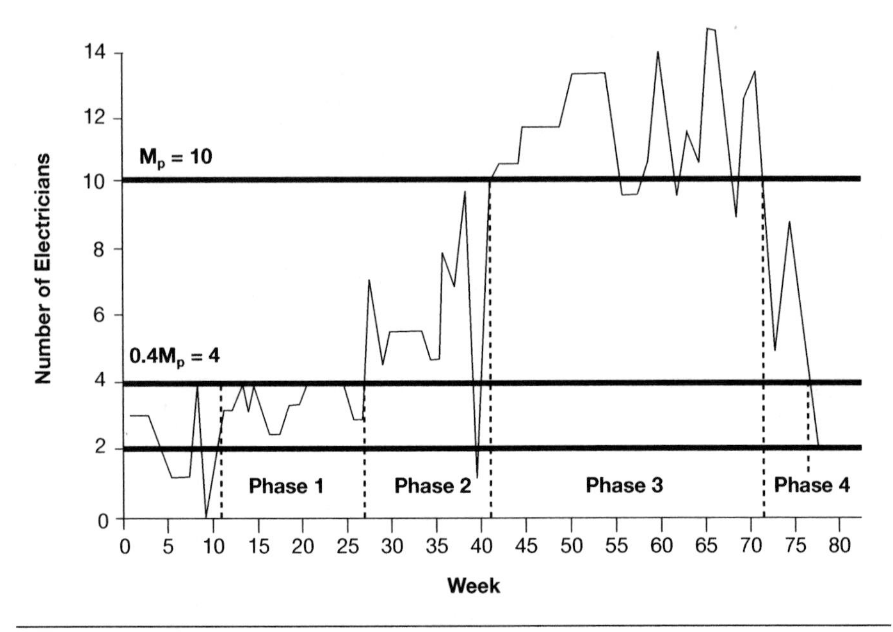

Figure 4.5 Project phases for the example project.

For example, a significant slowdown may result in going from Phase 1 to Phase 2 and back to Phase 1 again.

However, short periods of time should be ignored, as the consistency of the manpower level is what is important. Insignificant fluctuations beyond the defining limits are seen in Phases 2 and 3 of Figure 4.5.

Step 5: Determine the Weekly Performance Ratio

The patterns of labor inefficiency vary according to the phase, work conditions, schedule status, and the amount of work available to assign. The relationship between labor efficiency or PR and the labor rate deviation for Phases 1 to 4 is shown in Figures 4.6 to 4.9. To use these curves, we review the work conditions and inefficiency factors to ensure that it is appropriate to apply the curve. Not all conditions need to be present in order to use the curve. Also, the acceleration may begin during the middle of a phase, meaning that the inefficiency curve should not be applied throughout the entire time frame. Weeks when fewer than two electricians are assigned to the project should be overlooked.

Once the applicability of the curve has been established, the performance ratio for each week can be determined. We enter the curve with the labor rate deviation and determine the value of PR. Using week 48 in Table 4.4, the labor rate deviation is 0.37. Using the Phase 3 curve, the PR value is determined to be 1.25.

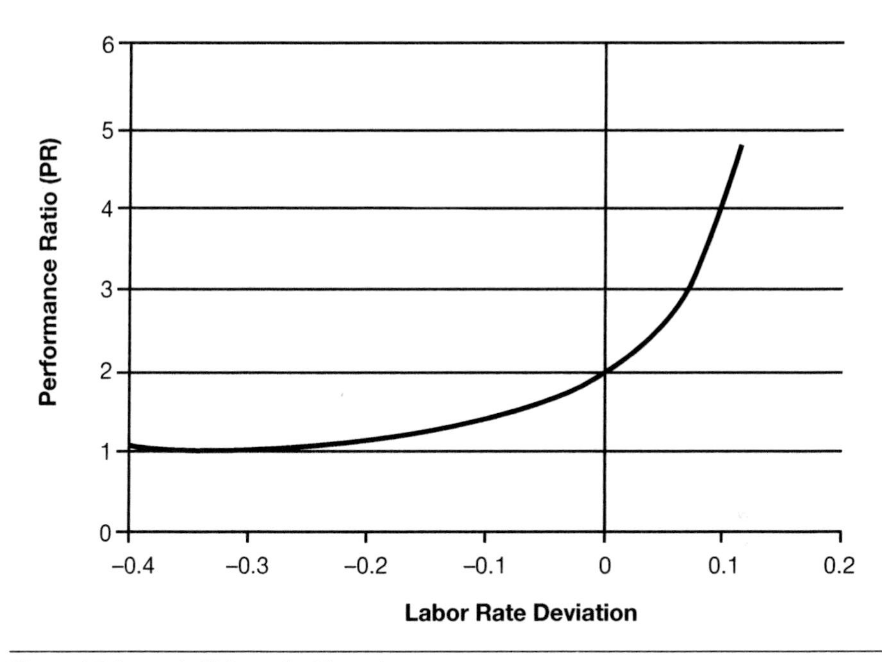

Figure 4.6 Loss of efficiency for Phase 1.

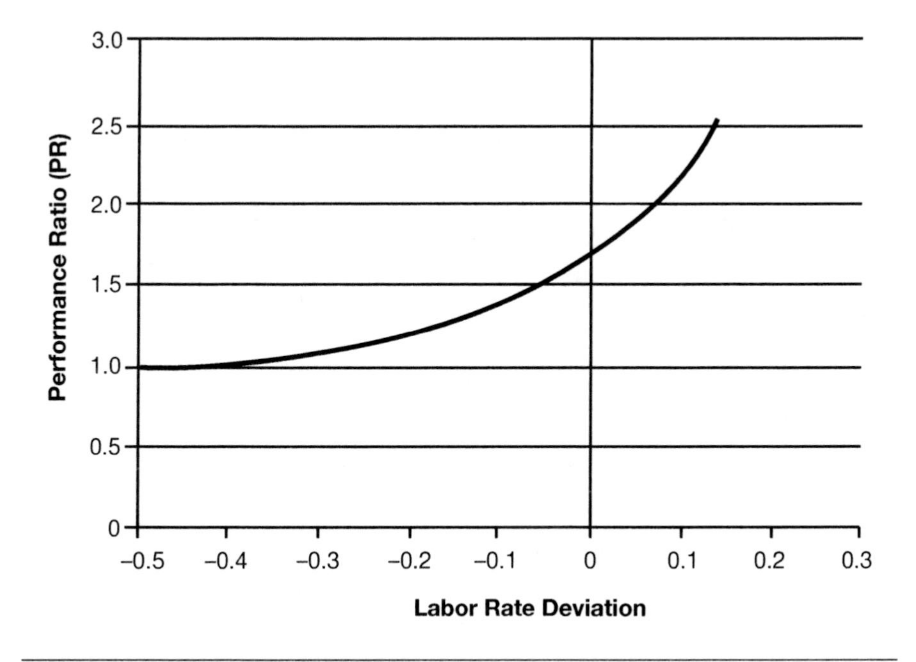

Figure 4.7 Loss of efficiency for Phase 2.

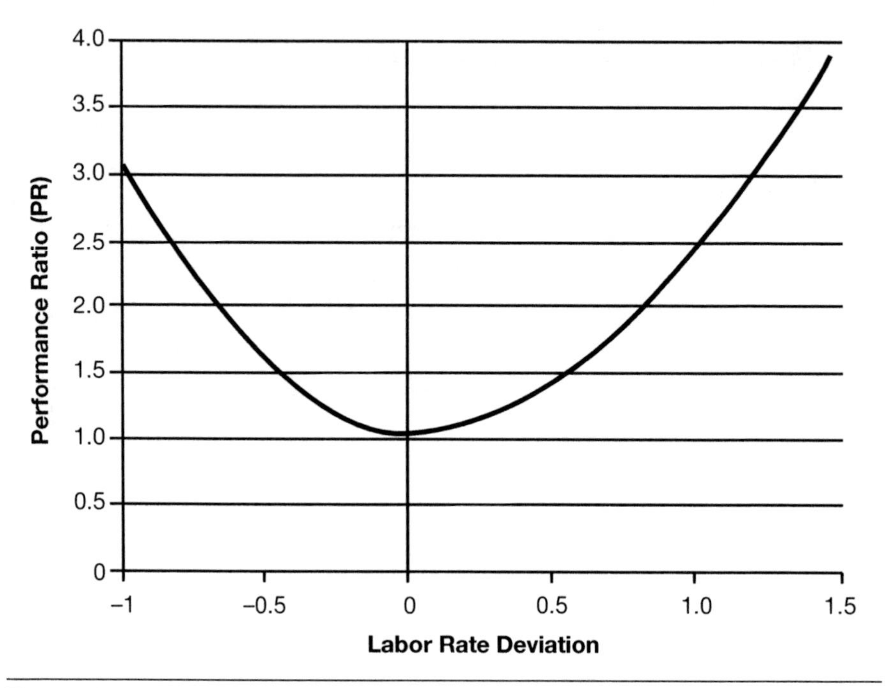

Figure 4.8 Loss of efficiency for Phase 3.

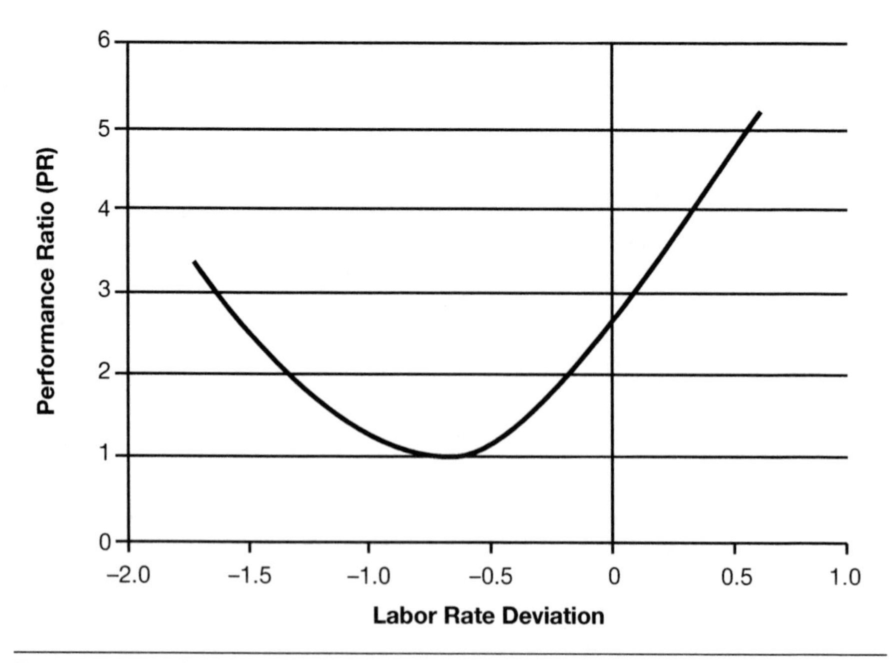

Figure 4.9 Loss of efficiency for Phase 4.

What does this mean? The PR value shows the ratio of labor hours needed to perform work that would normally be done at a ratio of 1.0.

Step 6: Calculate the Gross Weekly Inefficient Work Hours

For week 48, a total of 504 work hours were charged to the project. Using this value, the gross weekly inefficient work hours are calculated using the following equation:

$$\text{Gross inefficient work hours} = \text{Actual work hours} - \frac{\text{Actual work hours}}{\text{Performance ratio (PR)}} \quad (4.3)$$

For week 48, the gross inefficient work hours percentage is calculated as follows:

$$\text{Weekly loss of efficiency (\%)} = \left(1.0 - \frac{1.0}{1.25}\right)(100) = 20\%$$

Step 7: Calculate the Net Weekly Inefficient Work Hours

The inefficiency curves shown in Figures 4.6 to 4.9 were developed using several different types of projects. Therefore, it is necessary to adjust the gross inefficient work hours from Step 6 to reflect the specific project type. To convert the gross inefficient work hours to net inefficient work hours, the gross values are multiplied by efficiency loss multipliers. The multipliers are shown in Table 4.5. Since the example project is institutional, the efficiency loss multiplier is 0.82 and is applied as shown in Table 4.6. This table also shows the values for the entire project. Please keep in mind that the values for the entire project are not equal to the sum of the values shown in Table 4.6, since there are many more weeks in the project. Weeks 45 to 50 are shown in Table 4.6 for illustrative purposes. Table 4.6 also shows values for weekly loss of efficiency in percentage. This value is calculated using the following equation:

$$\text{Weekly loss of efficiency (\%)} = \left(1.0 - \frac{1.0}{\text{Weekly performance ratio}}\right)(100) \quad (4.4)$$

For week 48, the loss of efficiency percentage is as follows:

$$\text{Weekly loss of efficiency (\%)} = \left(1.0 - \frac{1.0}{1.25}\right)(100) = 20\%$$

Overall, the example project experienced a loss of efficiency of 25%, resulting in 3,812 inefficient work hours. That is, the example project took 25% more work

Table 4.5 Efficiency Loss Multipliers

Project Type	Efficiency Loss Multiplier
Industrial	1.05
Commercial	0.92
Institutional	0.82
Other	1.00

Table 4.6 Net Inefficient Work Hours

Week	Actual Work Hours (Hrs)	Performance Ratio	Gross Inefficient Work Hours (Hrs)	Efficiency Loss Multiplier	Net Inefficient Work Hours (Hrs)	Weekly Loss of Efficiency (%)
45	335	1.20	56	0.82	46	17
46	352	1.18	54	0.82	44	15
47	432	1.00	0	0.82	0	0
48	504	1.25	101	0.82	83	20
49	408	1.02	8	0.82	7	2
50	392	1.10	36	0.82	29	9
Entire Project	18,781		4,649		3,812	25

hours than it should have taken. The total project loss of efficiency percentage is calculated using the following equation:

$$\text{Project loss of efficiency (\%)} = \left(\frac{\text{Inefficient hours}}{\text{Actual hours} - \text{Inefficient hours}}\right)(100) \qquad (4.5)$$

Step 8: Validate the Analysis

The total labor inefficient work hours and the loss-of-efficiency percentage should be reviewed to ensure that they are reasonable. Unreasonably large estimates of work hour losses will likely forestall owner cooperation, and negotiations will be more difficult.

Labor inefficiencies do not occur at random but rather are caused by specific events or conditions, and the results need to be validated. We plot the weekly inefficient work hours as shown in Figure 4.10 and the weekly loss-of-efficiency

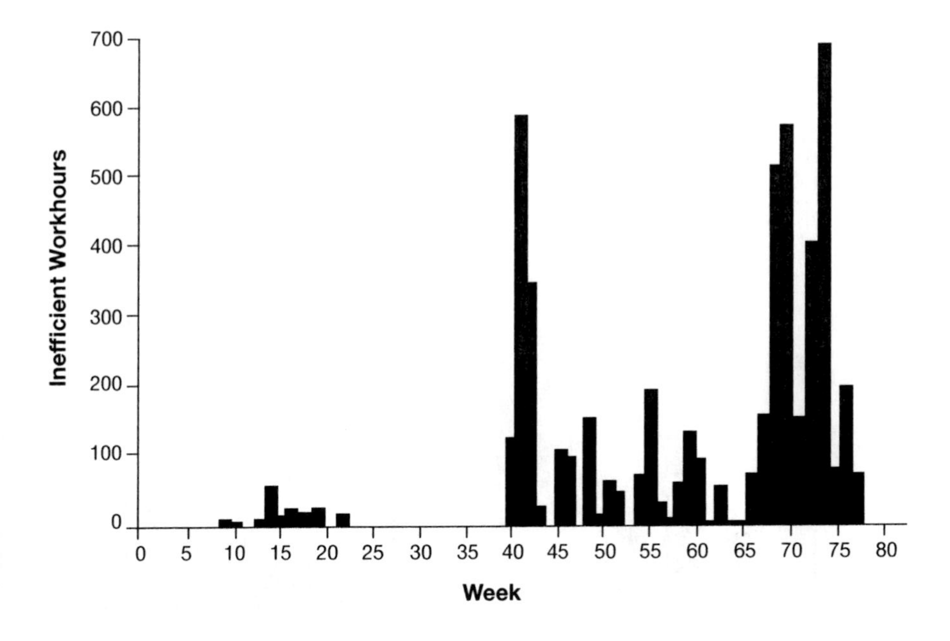

Figure 4.10 Weekly inefficient work hours for example project.

percentage as shown in Figure 4.11. Then we develop a chronology of events occurring in the project using Table 4.7 as a guide. As can be seen, for the example project, the inefficient work hours and loss of efficiency by week correlate reasonably well with the events leading to loss of efficiency. This analysis provides the basis for negotiation in that once the responsibility for an event is established the inefficient work hours resulting from the event can be estimated.

The methodology presented in this chapter accounts for labor inefficiencies caused by variations in the labor resources assigned to the project. The primary factors accounted for are the number of electricians assigned to the project and the work schedule, since these are the only ways the labor percentage can be increased or decreased. It follows that there are other factors leading to inefficiencies that may add to the total labor inefficiencies that are compensable. Following are some of these factors:

- Adverse weather
- Labor slowdowns and jurisdictional disputes
- Major changes
- Accidents
- Startup and testing
- Work in an operating environment

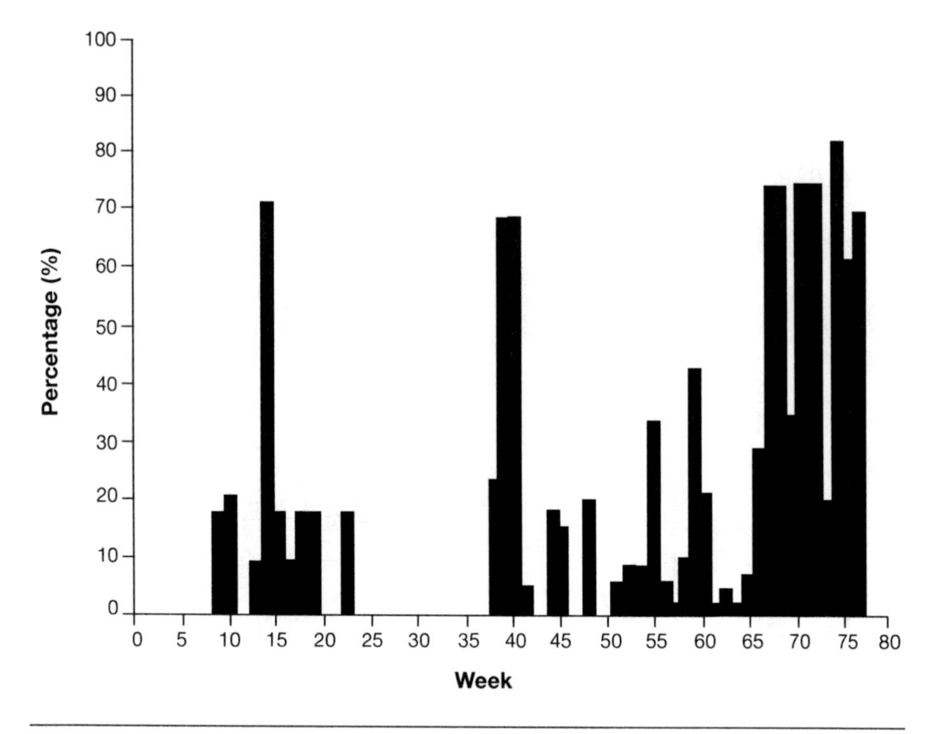

Figure 4.11 Weekly inefficient percentages for example project.

CONCLUSIONS

This chapter focused on how schedule acceleration and compression impacts labor productivity of electrical contractors. Only data from electrical contractors were collected when developing this approach. Therefore, the performance ratios developed for this methodology apply only to electrical contractors. However, the principles of schedule acceleration and compression apply to other specialty areas as well. Mechanical, sheet metal, piping, and other specialty contractors may experience similar consequences when facing schedule acceleration and compression and may want to calculate the effects of schedule acceleration and compression on their productivity. If this is the case, they should use the information in this chapter with caution and only to estimate ballpark figures of the possible impacts. Proper estimation of the effects of schedule acceleration and compression on trades other than electrical would require the replication of this study so that those trades develop their own versions of Figures 4.6 to 4.9 in order to calculate their own performance ratios.

Table 4.7 Chronology of Events on Example Project

Weeks	Events or Conditions
14–35	The work began slowly, in part due to steel fabrication errors, questions regarding rough-in service locations, and slow turnaround of submittals by the designer.
40	Major electrical and mechanical change issued requiring significant amounts of demolition. The amount of the change was about 25% of the total value of the two subcontracts.
44–55	Numerous errors and discrepancies with the electrical change order are found leading to periods of acceleration and deceleration.
56–62	The work slows considerably because of multiple problems. Most important, the casework subcontractor makes excessive contract demands and refuses to integrate his work with the other subcontractors. Chimney problems mean the roof cannot be finished and the work must proceed in very bad winter weather.
63–65	Continued delays waiting for the casework contractor.
66–72	Intense acceleration including overmanning and overtime to finish the work. Most of the project work has been compressed into an 11-month time frame rather than the 18 months allowed by the contract. The owner refuses to grant a time extension.

5

MANAGING SCHEDULE ACCELERATION AND COMPRESSION

Dr. H. Randolph Thomas, *The Pennsylvania State University*
Dr. Amr Oloufa, *The Pennsylvania State University*
Dr. Awad Hanna, *University of Wisconsin-Madison*
David A. Noyce, *University of Wisconsin-Madison*

INTRODUCTION

The purpose of this chapter is to provide guidance for managing schedule acceleration and compression to minimize impacts on construction projects. The basic principles of schedule acceleration and compression were studied in Chapter 4. When confronted with an acceleration and compression situation, contractors have three primary options for increasing labor hours: overtime, overmanning, and shift work. But increasing the quantity of labor hours does not automatically increase productivity. In some cases, productivity can actually decrease. Increasing labor hours while maintaining productivity depends not only on proper labor management but also on managing the resources that support direct labor at the worksite: materials, equipment and tools, information, and support services.

This chapter elaborates on the material presented in Chapter 4 to assist the contractor with managing schedule acceleration and compression. This management

WHAT IS THE PREFERRED ACCELERATION OPTION?

INSTRUCTIONS: Place an "X" in the column (option) as per the instructions given below. Not all issues are weighted equally, and managers and superintendents must weigh each issue against specific job conditions.

Schedule Overtime	Hire More Craftspeople	Additional Shift(s)	Description of Issues and Instructions
☐	☐	☐	**Availability of Good Supervision.** Place an "X" in the overtime column if there is limited supervision with experience at successfully managing acceleration. The other two options will likely require more supervisors and should be checked if more are available. In general, another shift is the most difficult situation. Multiple options may be checked.
☐	☐	☐	**Work force Availability.** If there are limited numbers of qualified and motivated craftspeople at the hiring hall, place an "X" for overtime as the preferred option; otherwise, leave blank.
☐	☐	☐	**Contract Language.** If the contract language states that the owner or prime will pay the premium time for overtime pay, place an "X" in the overtime column; otherwise, leave blank.
☐	☐	☐	**Overtime and Shift Differential.** If the labor agreement is particularly burdensome for either option, place an "X" in the most favorable option(s).
☐	☐	☐	**Congestion.** If adding more craftspeople will create a congested work environment, place an "X" in the overtime option. Also place an "X" in the multiple shift column.
☐	☐	☐	**Length of Acceleration.** If a short-term acceleration is anticipated, place an "X" in the overtime option. If the duration will be lengthy, place an "X" in the more craftspeople and multiple shift column.
☐	☐	☐	**Availability of Suitable Work for More Shifts.** If suitable work is available for more shifts, place an "X" in the multiple shift column; otherwise, place an "X" in the other two columns.

Figure 5.1 Preferred acceleration option.

has two major phases: (1) Completing the work—Completing the contract work in an accelerating fashion, despite the many problems that threaten to delay construction, requires maximizing labor efficiency and (2) Getting paid—One schedule acceleration principle seems constant: schedule acceleration and compression cost contractors more money, in additional labor hours, increased material handling and equipment costs, and related factors. Being fairly reimbursed for this additional cost is often largely a matter of careful documentation, with some up-front negotiation to clarify changed conditions.

This chapter details preliminary steps to be taken prior to the acceleration, both to maximize efficiency and to protect the contractor's interests. It also provides specific techniques for minimizing the impact of acceleration by maximizing productivity. The material presented in this chapter should assist the contractor in minimizing the impact of the accelerated schedule and provide guidance for the recovery of additional costs.

SELECTING THE LEAST DISRUPTIVE LABOR OPTION

In Chapter 4 we learned that when confronted with the need to increase weekly output, contractors have three basic choices: overtime, overmanning, and shift work. The advantages and disadvantages of each method are summarized in Table 4.1. In choosing which options to use, contractors should also review the contract language regarding rights associated with acceleration. In some contracts, the owner may not have the right to accelerate the work. It is not unusual for the owner to reserve the right to accelerate, but the owner may agree to pay the premium time associated with scheduled overtime. Figure 5.1 shows a form that contractors can use to determine the preferred acceleration options. Selecting the least disruptive option for acceleration will minimize adverse consequences.

NEGOTIATING WITH THE OWNER TO MITIGATE CONSTRAINTS

In Chapter 4, we learned that when alerted to an acceleration situation, contractors can take proactive steps to minimize the economic impact. The reasons for falling behind schedule need to be clearly identified and brought to the owner's or general contractor's attention in an atmosphere of cooperation. The constraints creating the schedule difficulty need to be removed or minimized. Otherwise, scheduling overtime and overmanning will not improve the situation.

A clear understanding of what is happening is the key to identifying constraints. Table 4.2 provides a brief synopsis of some of the possible constraints inhibiting progress and should serve as a guide. Identifying constraints should be done in conjunction with the owner and general contractor. Discussions should be frank and positive; a threatening or accusatory manner will not likely lead to a favorable outcome.

Unless the constraints causing schedule deterioration are removed, the schedule acceleration will become very costly. Contractors should obtain the cooperation of the owner and prime contractor if successful strategies are to be implemented. Communication and negotiation are keys to success. The following list includes some of the discussion items recommended for fruitful negotiation:

- Submittal schedule with dates for return to the contractor
- Dates for owner-supplied equipment, possibly adding more owner-supplied equipment
- Schedule for commissioning of equipment
- Joint funding of training program for new hires
- Revised responsibilities for trash removal
- Hosting by others and perhaps revised hosting times
- Partial relief from burdensome change approval process
- Change order work for time and material
- Owner payment of bonded warehouse lease so items can be purchased in bulk
- Relief from rigid specifications requiring restrictive material use and workmanship

MAINTAINING PROPER DOCUMENTATION

Many contractors fail to recover additional costs due to schedule acceleration and compression because of inadequate documentation. An invoice showing additional charges is not sufficient. There must be a showing of entitlement—that is, a cause-effect relationship showing that the increased costs were attributable to some action or event to which the written contract grants relief. What is often missing are the facts. Exact dates, meeting participants, content of conversations and directives, and a variety of other events may be forgotten.

Prior to construction, contractors must review contract clauses and addenda, especially procedures for notice, approval of substitutions, disputed work, changes, warranty clauses, and so on. The specific procedures should be clearly documented. Many instances have occurred where contractors failed to recover legitimate costs simply because the procedures clearly stated in the contract were not followed.

At the outset, it is advisable to clearly establish lines of authority. Specifically, the persons authorized to make changes and to whom notice should be given should be documented in writing. This step should be jointly done with the owner, designer, construction manager, and prime contractor.

It is widely recognized that the documentation practices on many construction projects are inadequate, untimely, or imprecise; that is, documentation frequently lacks critical dates and factual information. In particular, written communications often are not maintained properly, perhaps because of the urgency of the decision-making process. A party will have a difficult time prevailing in a dispute if that party is unable to substantiate the facts. Careful attention to documentation will permit speedy retrieval of vital information should a dispute arise.

Daily Reports and Logs

Often the daily report is not much more than a manning report or head count. Daily reports make better documentation to support claims for extras when the work is divided into tasks and work areas. In this way, the daily reports can track the number of hours per area per activity and crew movements. If appropriate, new report forms should be created to capture this information.

Daily crew quantity reports are invaluable for substantiating losses of efficiency. These reports need not be complicated and should be designed so they can be completed by a crew foreman in no more than several minutes. A report should be required from each crew, noting significant problems or interferences affecting the crew's work. Further, it is very important that there be a record of activity prior to the events causing the inefficiencies for which the contractor is seeking recovery. Such information can be used to establish a baseline or "measured mile."

Another valuable source of information is logs. It is useful to maintain logs of submittals, RFIs, change requests, and so on. Such logs should include dates so that with little effort, contractors can substantiate the time elapsed waiting for certain actions and how this turnaround has changed over time.

Estimates of Productivity Losses

At least once a week, the site superintendent should calculate or estimate work hours lost that prior week due to double-handling of materials, mobilization/demobilization cycles, stop work areas, and any other important reasons. The quantity reports described earlier can be used as one basis for this estimate.

Walk-Throughs

Conduct regular walk-throughs (once or twice a week), with an audio recorder, and record disruptions, stop-and-start work, and incomplete predecessor work; then transcribe this information in narrative form. Although the recorded facts may be somewhat random, the narrative creates a descriptive record of job interferences.

Photographs and Videos

Photographs and videos are valuable in explaining problems and their impacts on a project, especially where the presentation is made to someone not familiar with the job (mediator, arbitrator). Use a camera that automatically records the time and date on the photo or video. Photos and videos should record progress and job status, document problems, and show interferences. Be liberal with taking photographs and videos.

Notice Letters

Develop form letters with a first and last paragraph and blank space in between. Whether the information on delays or interferences is handwritten or typed, forms make it easier for the site superintendent to satisfy the contract written notice requirements.

Every contract has written notice requirements, and some have fairly burdensome demands. Make a checklist for the timing and amount of detail needed to preserve rights to equitable adjustments for delays, extra work, changes, suspensions, etc., and post it in the field trailer.

Each occurrence of a change or delay should have a corresponding written notice letter, although continuation of a previous notice condition does not necessarily require another notice letter. However, in providing written notice, it is important that the right to additional claims be preserved. Standard language reserving all contract rights should be clearly presented in all communications.

Construction schedules can sometimes demonstrate notice and are frequently used in delay claims. The prudent contractor should not overlook this possible application. In using the schedule as a notice document, the following requirements should be met:

- Updates must highlight delays.
- Responsibility for the delay should be indicated.
- The schedule should not contain errors.
- Documentary evidence of the delay should be submitted with the schedule when appropriate.

Tracking Schedule Progress

Regardless of how simple the original schedule, create a detailed bar chart and post it in the job trailer. Each day, have the general foreman color-code actual progress against the plan. If possible, man-load and equipment-load the plan, and then track actual manpower and equipment on the chart. For larger jobs, bar charts may need to be developed for specific areas of the project.

Changes File

Too often, the effects of extra work on performance, particularly during acceleration, cannot be established in proving a claim. If possible, create unique codes on the daily report, daily quantity report, and timesheets for each extra, even if it has not been approved. Be sure to track work hours. As a minimum, note when each extra work activity starts and when it is completed. Even for time-and-materials work, courts often require detailed records to establish actual costs. This information is also useful in determining the effect of changes on unchanged work.

A relatively simple filing system for handling all correspondence related to changes can reduce the confusion of tracking the progress of a change. Each change should generate a separate file for correspondence, drawings and specification revisions, and cost estimates. Where the schedule has been affected, the original and revised sequences of operations should be carefully documented. Maintenance of a separate file for each change will provide that revisions to changes can be easily included, all the relevant information on a change is accessible at one time, and cross-referencing of related materials between the self-contained files will be facilitated.

Minutes of Meetings

Minutes of meetings can be an effective form of communication; however, most minutes contain insufficient information and do not cite pertinent contract provisions. Minutes should always be distributed in a timely manner, and submission of any corrections should be required by a specific date. Failure to respond indicates acceptance of the minutes as being factually correct and may effectively bar contrary arguments at a later time. Keeping minutes is often viewed as an onerous or useless task, but a prudent contractor would be wise to do so. Keep your own minutes, even if not assigned to do so. Even informal meetings should be documented by minutes.

Content of Communications

Letters and memorandums should be used to accurately communicate factual information. It is recommended that the content be limited to a single issue. Correspondence should include the following:

- The date of the correspondence.
- A detailed description of the conditions that the correspondence addresses.
- The rights established in the contract.
- Reference to applicable contract clauses.

- The intent to actively pursue a solution to the problem.
- A statement that additional compensation is expected as provided for in the contract.
- The provision of a reasonable response time if action by the recipient is deemed necessary.

Proper information is not always communicated. For example, letters often describe difficulties and complaints but fail to relate the problem to a specific contract provision that entitles the contractor to additional compensation. Criticisms of other parties should be avoided, and legal terms and connotations should be left to attorneys.

Use of Consultants and Attorneys

Expert advice and testimony is frequently used in negotiations and in court for judgments as to the effects of interferences and acceleration on labor productivity. This testimony will be weakened substantially if the expert has no firsthand knowledge of the problems. On the other hand, the experts' opinions have greater influence when the expert can state that the opinions are based on observations made while the work was in progress.

The typical contract is filled with disclaimer clauses and other contract terms that make the contractor or owner not liable under specified circumstances. There are no damages for delay clauses, cooperation among contractors clauses, and written notice provisions for delays, extras, and so on. Contractors should consult with legal counsel to determine how they can best use the contract language and how they can best protect themselves against inadvertently waiving a claim. The attorney could assist in developing the form letters that satisfy the contract's notice requirements.

Table 5.1 summarizes the typical documentation that prudent contractors should maintain to protect themselves in case a dispute arises regarding schedule acceleration and compression.

STRATEGIES FOR MITIGATING PLANNED SCHEDULE ACCELERATION

There are many steps that contractors can take to minimize the impact of an accelerated schedule. Sometimes schedule acceleration is planned well in advance, based on external causes such as the owner's or tenant's need to occupy the structure well in advance of the original contract date. More often, schedule acceleration is not planned in advance but is caused by jobsite factors such as poor weather,

Table 5.1 Proper Documentation for Schedule Acceleration and Compression

Item	Recommended Documentation
Preconstruction	Contract procedures for notice, submittals, RFIs, change requests, etc. Lines of authority for owner, designer, construction manager, prime contractor, etc.
Daily Reports and Logs	Daily reports showing hours per area and per activity and crew movements Daily crew quantity reports denoting problems and interferences Logs of submittals, RFIs, change requests, etc.
Loss of Productivity Estimates	Weekly estimate of the work hours lost due to specifically defined causes
Photographs and Videos	Photographic and video history of the job, with particular emphasis on showing progress and documenting problems and interferences
Notice Letters	Checklist of timing and detail requirements of the contract notice requirements Written correspondence for each occurrence, citing the specifics of the problem, applicable contract clause permitting cost or time adjustments, and language preserving the right to additional claims Accurate and timely schedules updates showing the effects of delays and responsibility
Schedule Progress	Bar chart of planned and actual progress
Changes File	Unique codes on all reports to track quantities and work hours for each change Separate file for each change containing correspondence, drawings, specification revisions, cost estimates, interferences, etc.
Meeting Minutes	Minutes of meetings, even informal meetings
Letters and Memorandums	Correspondence to responsible individuals describing specific factual information

approval delays, or delays caused by other trades. In many instances, the techniques used by contractors are similar in both cases. While planned acceleration allows more advanced planning and preparation, schedule acceleration and compression is characterized by rapidly changing conditions. In other words, a contractor is often reacting to changing conditions, no matter how much advanced preparation came before. During unplanned acceleration, and with careful attention to changing conditions and management techniques, a contractor will become more skilled at dealing with schedule acceleration and compression the longer the situation goes on.

The bottom line is that the basic conditions of planned and unplanned schedule acceleration and compression are much the same. Over time, they begin to resemble each other even more. This is why many of the same techniques are used in each case. The concepts included in this section are primarily planning and scheduling techniques that can be employed when the schedule acceleration is known in advance. The concepts included in the next section ("Unplanned Schedule Acceleration") are day-to-day management techniques that may be useful in either a planned or unplanned compression situation.

Use Critical Path Method Scheduling

Using detailed scheduling techniques such as critical path method (CPM) in compression situations clearly defines required activities and scheduled durations. This allows the contractor to focus efforts upon items on the critical path. Since most construction claims contain some aspect of delay, establishing a formal CPM schedule at the beginning of the project assists the contractors in sequencing and showing work effort. CPM and other detailed scheduling techniques should be used on all projects with preplanned accelerated schedules. This will probably require a full-time staff planner or consultant. Rapid changes on an accelerated or compressed project will required frequent schedule changes and updates. The additional cost of a planner may be offset by more efficient use of resources including labor, equipment and tools, and materials.

Reduce Task Scopes to Milestone Activities

Breaking down larger tasks into small milestone activities gives the contractor more flexibility in reacting to changes. Smaller tasks allow crews to complete tasks while avoiding the remobilization and change work typically associated with schedule changes. Additional milestones allow the contractor to better control the project schedule and costs. This concept becomes more important as schedules become more compressed. The project must be large enough that smaller tasks still have sufficient work components to maintain crew productivity. Management must plan milestones and task scopes sufficiently in advance to ensure that materials, tools, and successor tasks are ready when needed. Smaller tasks allow for better cost control under the dynamic environment of schedule compression.

Include Anticipated Weather Delays in Scheduling

Almost all construction projects involving outdoor work will face some delay because of weather. Allowing for this contingency in the advanced planning minimizes disruption of an accelerated schedule. This concept is more important in the early stages of a project, until more inside work becomes available. Potential weather delays should be factored into all outdoor projects with preplanned accelerated schedules. This weather scheduling should apply to all trades. Allowing for weather delays in an accelerated project may help the contractor avoid penalties and liquidated damages.

Assign a Material Control Coordinator to the Project

As project schedules accelerate, the flow of materials to the site increases, as do changes in the materials required. A project material coordinator works to ensure that crew productivity does not suffer due to lack of needed materials. The material coordinator is also responsible for ensuring that cranes, trucks, and forklifts are available to transport materials on-site. Material coordinators should be used on large projects under a compressed or accelerated schedule, or when material storage areas are limited and material flow is critical for project success. The material coordinator must be knowledgeable about all aspects of the project and understand the needs of the production crews. This requires adding a supervisory position to the project, though the additional cost should be offset by increased productivity.

Establish Material Lay-Down Areas

Organize material lay-down areas to match construction areas to allow easy access for delivery vehicles, minimize travel distance from delivery point to point of use, and minimize double-handling of materials. This requires secure storage areas at the jobsite such as a movable fence with locked gate, storage shed, or trailer. There should be sufficient space for shakeout of materials (layout in sequence needed) and on-site prefabrication of assemblies.

Establishing material lay-down areas is most useful on large projects that are spread out and have a large number of crews on-site. This must be planned in advance with all other trades and contractors. The project must have ample space for on-site storage of construction materials. Due to the constant changes characterizing an accelerated job, it may be difficult to maintain an elaborate material storage and delivery system over the life of the project. The costs of developing a secure storage facility on-site may be offset by the savings in material handling and storage costs.

Employ Special Shift Arrangements

This concept involves the use of four 10-hour days, variable shift lengths, and other innovative schedules. Workers tend to adjust to 10-hour days, and productivity is maintained in the later hours. Using special shift arrangements can increase production while improving productivity. The labor agreement must allow 10-hour days, and the owner or general contractor must agree to these special schedules. Rolling four 10-hour days are best suited for long-duration projects and highly repetitive work that is engineered and detailed beforehand. Special schedules must become routine to be successful. They generally are not suited for short-term accelerations. Owners and contractors tend to dislike variable work schedules. Setup and miscellaneous work can be done by a small crew on Fridays, or Fridays can be used as makeup workdays without paying overtime.

Use Modular and Preassembled Components

Move as much work as possible off-site or to areas away from the worksite. Storage areas and buildings can be used to set up mini assembly lines or prefabrication facilities for small items. Repetitive volume work can be done away from interference, noise, and weather problems, and without contributing to jobsite congestion. This concept works best when large numbers of components can be preassembled. Apprentices can be used for the assembly operation. Productivity is enhanced by working in a favorable indoor environment.

STRATEGIES FOR MITIGATING UNPLANNED SCHEDULE ACCELERATION

As mentioned, the concepts included in this section are day-to-day management techniques that may be useful in either a planned or unplanned compression situation.

Monitor Progress Daily or Weekly Instead of Biweekly or Monthly

Because the work is progressing faster, more frequent monitoring of installation rates, productivity, and costs are justified. A cost-effective way to monitor work is manually, exclusive of the computerized cost accounting system. The monitoring may selectively concentrate on the most important quantity items that will control the schedule. Monitoring can be done by task, system, interface area, or isolated area. More frequent monitoring can rely on input from the job foreman. It is also valuable to develop a planned or baseline manpower curve. A daily diary of problems should be maintained. The diary should cover significant events; visitors to the site; communications with the owner, designer, general contractor, and other key individuals; and weather conditions. Be sure to include items that were helpful to productivity.

Where possible, baseline productivity rates should be established that verify the productivity and production rates before the acceleration. To establish the before-acceleration rate, contractors need to anticipate the acceleration situation;

otherwise, the opportunity will be lost. A productivity curve shown to the owner can be a valuable document from which to negotiate. Manual monitoring can be applied at any time, even when there is no acceleration. An efficient computerized system is important if the cost monitoring system is to be used to track progress. To rely on the foreman, the measurement scheme needs to be limited to relatively simple counting. Do not try to manually measure too many activities.

The cost implications are very small. Manual monitoring can be efficient so long as foremen and superintendents are not overloaded with data. Some staff time will be needed to develop the baseline manpower curve.

Do More Detailed Planning and Short-Interval Scheduling (SIS)

One of the more serious consequences of an accelerated schedule is the inability to plan very far ahead. Foremen and supervisors should insist on developing short-range (one- to three-day) schedules. Contingency plans should also be developed should the schedule change; keep several options available. Planning should include reconciling and studying the drawings and identifying all the resources needed.

The traditional short-range schedule developed in industrial engineering focuses on how much work can be done in a given time frame. The schedules used by contractors should focus on how long it will take to complete a given amount of work. The more accelerated the schedule, the greater the need for the short-interval schedule. A clear indicator of this need is having to do out-of-sequence work. The planning horizon should be progressively shorter as the schedule becomes more compressed. Contingency plans need to be a part of the planning process. The major cost implication of short-interval scheduling is that it may require some training of crew foremen. More time may be spent by supervisors in emphasizing the schedule and contingency plan.

Improve Communications between Crews

Craftspeople have many good ideas, and supervisors should take advantage of their expertise. Foremen often have informal communication networks, and they may be able to get things done with limited interaction with the superintendent. When asked, craftspeople are usually very willing to share their problems and frustrations. Planned meetings—say, weekly—can be useful. When working multiple shifts, supervision should always overlap, and daily turnover meetings should be the norm. Crew briefings relative to safety means that supervisors and foremen should plan the short-term work before talking with the crew.

Good communication among crews is helpful at any time. However, the acceleration is an opportunity to get the project manager's, superintendent's, and foremen's attention to do many things that should have been done as a matter of routine. Well-planned meetings are essential. Generally, one planned effort at improving communications is the only opportunity available to seek input from the crafts; therefore, superintendents should plan this concept carefully. Craftspeople will respond favorably when they see their problems are being resolved and their job frustrations are being addressed. Listening and not acting may cause more harm than good. It may be necessary to meet after work hours, thus necessitating overtime pay. Turnover meetings will involve additional premium time wages for supervision. Otherwise, the cost implications are minimal.

Improve Communications with Other Contractors, Construction Manager, and Owner

Communicate regularly and informally with other contractors. When a cooperative attitude is shown, others usually reciprocate. At the manager level, more coordination meetings may be needed during acceleration. Meetings should be conducted in as positive an atmosphere as possible. Accurate and timely minutes are very important. When minutes are published, they should be read carefully, as they will become the record of what happened on the job. Errors or deletions should be promptly brought to the originator's attention. It will be difficult to argue later that a particular item was or was not discussed.

Meetings will become more frequent as the work progress falls behind schedule. It may even be prudent to attend or participate in meetings before arriving on-site to begin work. The project manager and superintendent should strive to develop good relations with the owner and general contractor. Meetings may not be especially productive if they degrade into gripe sessions or shouting matches.

Monitor Absenteeism for Detrimental Trends

Absenteeism is an indicator of a poor project. Absenteeism levels will rise as site conditions deteriorate. Craftspeople that are habitually absent should be terminated. There should be a written policy that a craftsperson who is absent cannot work overtime in the same week. For planned reductions in force, begin with those who have the highest absenteeism rate. The person who is habitually absent is often the least likely to work steadily, produces the poorest-quality work, complains the most, and is the most likely to have safety violations. If there is a policy of not tolerating habitual absenteeism, reductions in force can be done at any time. There needs to be a good but simple tracking system so that reductions in force do not seem arbitrary. Reductions in force should be consistently done with a written policy on absenteeism. There is a cost for replacing a worker; however, this cost may be offset by a more productive workforce.

Add More Foremen and Supervisors as the Workforce Is Increased

A union agreement may specify the ratio of foremen and general foremen to number of craftspeople. However, when there is acceleration, more supervisors and nonmanual persons will probably be needed. Smaller crew sizes may be more responsive to the changes on the jobsite, thus increasing the need for more foremen.

The added supervision can be effectively used to perform nonsupervisory functions, including monitoring progress, material requisitioning and handling, expediting, reviewing change proposals, and studying plans to identify errors and ambiguities. Additional supervision is best applied when the specific work assignments have been identified. This concept has significant cost implications because adding supervision involves quantifiable cost expenditures. The savings in costs from adding supervision are very difficult to quantify.

Add Supervision to Identify Design Errors and Omissions

Information requests, drawing errors, and changes are a significant cause of schedule deterioration. When crews stop because of a design question or drawing error, losses of productivity for that work and any subsequent change can easily exceed 25%. Any effort to identify errors and ambiguities will pay big dividends in morale and productivity. The original scope of work and subsequent changes should be reconciled into a single scope of work by someone other than the crew foreman. This will free the foreman's time for supervising instead of the frustrating job of studying and reconciling drawings.

It may be cost-effective to add an estimator to the job to cost changes work. Negotiation with the owner and contractor on selected items may alleviate delay problems. Contractors can negotiate to streamline the approval process or to gain relief from rigid specifications that restrict material use. Jobs become seriously impacted when the work hours spent on changed work exceeds about 10% of the total work hours.

Contractors should routinely track hours spent in rework caused by poor workmanship. Obviously, there are cost implications to assigning another supervisory person to the project, but this cost involves a trade-off with the potential savings on labor productivity when the drawings have been reconciled before being assigned to the crew for work.

Deliver as Much Material to the Job as Practical

Since the schedule may change almost daily, foremen cannot be sure what work areas will be released next or when access to certain areas will be denied. Contingency planning dictates that to avoid crews having work available but no materials, much more material needs to be delivered than would be normally expected. Adding a material delivery crew may be highly desirable. This crew may be effectively used on a second shift. The superintendent should rethink the material delivery schedule. Price advantages may result from buying in bulk. Equally important is the removal of as much unneeded materials as practical off the project site.

Adequate storage space is an important consideration. Clearly, there needs to be adequate storage areas, and there should be some coordinated effort to develop a sitewide storage plan. Off-site storage space may be needed. Shipments may need to be diverted to the contractor's shop or yard or to a bonded warehouse. These strategies will lead to double-handling of materials.

The material deliveries need to be efficiently done. Materials need to be clearly marked and stored for easy retrieval. Damage and theft may increase. Bundling of items can reduce this problem. Contractor payment problems can also occur and should be a topic of negotiation with the owner. Depending on how materials are packaged and shipped, it may be necessary to rethink the equipment used for material handling. For example, pallet jacks may need to be rented.

Material management is a serious reason for losses of labor efficiency. Thus, any cost in streamlining this function may pay big dividends on improved productivity. There will likely be a need for more supervision. Supervision will be needed to reschedule deliveries, evaluate delivery strategies and equipment selection, and oversee the organization of the storage areas.

Reschedule Long Lead Items

Items with longer lead times need to be specifically identified because these items will probably need to be delivered sooner than originally planned. Larger equipment items may need protective measures, and smaller items will need protection from damage and theft. Anticipation of the acceleration is important. It may be worthwhile to negotiate with the owner and designer to substitute more common stock items and components.

Some rescheduling will involve coordination with the owner and contractor. Beware of holiday and summer vacation schedules, especially if ordering from international firms. Many firms are closed for as long as a month. Shortened delivery schedules will probably cost more.

Reassess Material Handling and Storage Areas

Because of frequent schedule changes, the areas adjacent to and within the facility may be required for other work on very short notice. Materials stored there will need to be moved, wasting valuable craft and equipment resources. What may have been a good storage area before acceleration may be too close during acceleration. Move material away from these chaotic areas to minimize the potential for double-handling. Store only one or two days' supply of materials in these areas. Longer-term storage may be more convenient in a warehouse or subcontractor's shop or yard.

Miscellaneous items like ladders, scaffolding, pallets, and so on should be removed to a remote location promptly after use, as these will interfere with other contractors. A fence or secure area may be needed. The handling and storage areas should be reevaluated on a very chaotic job. Shortness of delivery distances is normally a criterion, but it may not always be best when schedules are accelerated. Double-handling of material is a sure indicator that it is time to reevaluate storage locations. Provisions for rapid and efficient delivery of materials are probably needed in conjunction with this concept. When changes are needed, the contractor is reminded to maintain good documentation.

Cost savings from avoiding double-handling may offset some added cost caused by having longer delivery distances. Storage in a warehouse, especially if bonded, or in the subcontractor's shop or yard will add to the cost of material handling. Carefully kept cost records are important because contractors may be reluctant to charge enough for this disruption.

Organize a Special Material Handling Crew

A special crew for organizing and delivering materials allows production crews to begin work immediately and not waste time locating and gathering required materials. This crew may work a night shift to have materials in place for the regular production crews.

Excess material deliveries to the work station will lead to double-handling of materials, especially with frequent schedule changes. The size of the material package should probably be the amount of work that can be done in one or two days. All bundled and palletized materials should be clearly labeled and include an inventory for the foreman. The size of the delivery crew should be tailored to the amount of work available.

Negotiations with the owner or contractor may be fruitful. Negotiate for the use of more expensive but efficient forklifts and pallet jacks. Renting equipment may be necessary. Sharing hoisting equipment with another contractor may defray some of the added cost. Delivering materials at night when elevators are available may also be a worthwhile option. This concept will probably pay dividends at any time, even if the schedule is not accelerated. Containerized material packing is a good means of delivering and storing materials for specific work tasks.

There will be increased cost for the special crew, the package containers, and the delivery equipment. Rental of hoisting equipment will also be a cost factor. These costs should be partially offset by the productivity improvements of the work crews.

Organize a Setup/Preparation Crew

A specialized setup crew can organize and lay out the work for the day shift crew. This work can include gathering tools and equipment or erecting scaffolding, for example. Setup work may be best suited for a second or staggered shift, to avoid interfering with work crews. The size of the setup/preparation crew can be tailored to the amount of work. This concept is suitable for short-term accelerations and compressions. It works best when project tasks can be broken down into definable subtasks of small scope.

There must be sufficient preparation work to justify a special crew. The setup crew must have a full understanding of the work and what needs to be completed by the production crew. The overtime cost for the special crew may be offset by increased productivity of the regular work crews.

Organize a Special Cleanup Crew

Faster work causes waste materials and trash to accumulate at a faster rate. Poor housekeeping is a significant cause of lost productivity and a safety hazard. Therefore, it is important to address the need for expeditious trash removal. Organizing a special cleanup crew frees the production crews to concentrate on completing work assignments.

This work may be conducive to a second or staggered shift to avoid interfering with work crews. The size of the cleanup crew can be tailored to the amount of work. This concept is applied when a reasonable amount of waste material is produced. The crew will need a way to dispose of waste. This may require extra waste containers or roll-offs.

There could be union jurisdiction problems with this concept. Caution should be exercised that items to be installed do not get swept up and discarded in an overly aggressive cleanup campaign. There will be overtime costs for the crew. These costs are less than the cost of not keeping the work area clean. Consider sharing the cost and work with another contractor.

Organize a Special Crew to Handle Changes

A special crew to handle changes keeps production crews working and building momentum. The changes crew should not follow too closely to the production crews. It can be bad for morale if production crews see their work being torn out. If the cause for the change is poor workmanship, the production crew should probably be required to perform that work. Stop-and-go operations result in loss of momentum and poor morale. The impact to productivity can be quite significant.

Shift to Smaller Crews or Work Teams within a Crew

Schedule acceleration tends to lead to shorter planning horizons due to the potential for schedule changes. With smaller work packages, smaller work teams can perform this work. Also, if the work is unexpectedly halted, the disruption affects fewer craft workers if a smaller crew is used.

This concept is most often associated with the changeover to the package concept or the startup phase. Crowded work areas can be relieved somewhat with smaller crews. This concept is more attractive as the job progresses and the emphasis shifts from installing bulk quantities to completing specific components and systems. This typically occurs when the work is about 80% to 90% complete. More supervision may be required. Foremen and superintendents must control labor so a one-person job is being done by one person.

Small work teams place an extra burden on the foreman as the teams can be widely scattered about the facility. Communications can be a legitimate concern. Also, more and smaller work packages will need to be prepared. This can place an added burden on setup crews. Once bulk production has been largely completed, productivity rates for normal large crews begin to suffer, making a change to smaller crews more attractive.

Insist on Greater Backlog before Beginning Work

A serious detriment to productivity is not having sufficient work to do. The problem is compounded when working in teams because the idle team must locate the foreman for instructions, and this may be difficult. To alleviate this problem, resist the pressure to do piecemeal work and wait for a sufficient backlog of work to keep a team or crew busy for the total workday. This situation is sometimes associated with the changeover to the package concept or startup phase.

Stress Crews Perform 100% of their Work

Rework because of poor workmanship or just incomplete work is very detrimental to productivity. Supervisors should insist that the work be 100% complete before crews move to other work. Even a 99% complete work package may preclude follow-on crews from being able to do their work. Thus, incomplete work compounds the productivity problem. Consider developing a check sheet.

Incomplete work can occur when schedule impacts are acute because under this situation, there is a greater emphasis on completing critical activity work. This situation can become acute toward the end of the job. There are few cost implications to this concept.

Increase the Inventory of Spare Parts, Hand Tools, and Expendables

Increased tool use in an acceleration environment increases the need for tool maintenance and replacement. Studies have shown that worn or missing tools and expendables can reduce output by 25% and can have a negative effect on work quality.

A program of exchanging worn-out tools and expendables like drill bits and grinding wheels for newer ones may be beneficial. It may be worthwhile to add a tool-person to the job. In larger metropolitan areas, it may be possible to obtain half-day vendor service on broken and worn tools. Tool theft can become a serious problem.

Tools are needed at one tool per craftsperson, not per work hour. This adds to the cost as the workforce size is increased. More tools can add greatly to project cost. Cost savings may be realized on daily equipment rental rates if they can be used on a second shift.

Be Selective on the Work Assigned to a Second Shift

The second shift is not a copy of the first. Work assigned to the second shift should not require much engineering or design support. It should be able to be completed before the first shift returns; otherwise, there will be incomplete work left that adds to the clutter of the job.

Activities that will be easier to do in the relative quiet of a second shift are good candidates. Contractors should negotiate with the owner on a number of points related to the work. Items that are candidates for a second shift include moving materials, work requiring equipment or lifts, quantifiable tasks, corridor work, and work requiring hammering (hammer drills).

Use Modular or Preassembled Components

Move as much work as possible off-site or to areas outside the immediate vicinity of the facility. Storage areas may be a convenient area to set up mini assembly lines or prefabrication facilities. The work can be done away from interferences, noise, and congestion.

Use Special Shift Arrangements

This concept involves the use of rolling 4- to 10-hour days, variable length shifts, and other innovative schedules. Set up and miscellaneous work can be done by a small crew during off-peak hours. Fridays can be used as a makeup workday without having to pay overtime.

CONCLUSIONS

This chapter introduced specific strategies for managing schedule acceleration and compression. Any specialty or general contractor can benefit from the recommendations provided. In some instances, the recommended practices can be beneficial even without the presence of acceleration. Acceleration and compression is the response to schedule slippage. If a contractor is asked to accelerate, special consideration should be given to the potential constraints that can be the root causes of the schedule slippage. Removal of constraints is essential to eliminate the fuel that is feeding the problem. Since schedule acceleration and compression always involve additional costs to the contractor, open negotiations with the owner and general contractor are important to alleviate some of the additional burden and preserve the rights to present future claims. Proper documentation of the project environment before and during acceleration is necessary to prove the impact of acceleration on project performance. Finally, specific managerial approaches can be applied to minimize the adverse effects of schedule acceleration and compression. However, it is important to keep in mind that some of these techniques may involve additional costs.

6

ABSENTEEISM AND TURNOVER

Dr. Awad Hanna, *University of Wisconsin-Madison*

INTRODUCTION

Labor costs for contractors can reach as much as 40% to 60% of total construction costs. With that much invested, contractors need a workforce that is both stable and productive. Research has found that absenteeism and turnover reduce productivity. While this is not surprising, the construction industry has traditionally paid little attention to these problems. That might be because contractors believe these problems just come with the territory. But once contractors understand why the problems occur, they can make some changes that will decrease absenteeism and turnover, and that should lead to an increase in productivity.

Production suffers when even one worker is absent for just one day. Unfortunately, the problem is significantly worse than that. Absenteeism on some construction jobs has been reported as high as 20%. Voluntary turnover (workers quitting) is also common. Contractors need a way to figure out how serious these problems are. They also need solutions to help reduce their effect on productivity. The good news is that management can change this situation.

This chapter explores the potential causes for absenteeism and turnover, illustrates their consequences, and offers solutions that can help contractors improve productivity.

PREVIOUS STUDIES

Previous research on the construction industry has shown absenteeism rates at 20% and turnover rates at 200% annually. Yet scant research exists on the effect these problems have on the construction industry. This section presents an overview of previous studies on absenteeism and turnover in the construction and other industries.

In general, researchers agree that companies can divide the factors causing absenteeism and turnover into controllable and uncontrollable. Among controllable factors, we can cite management capabilities, other crew members and teamwork, excessive rework, and job satisfaction. Among uncontrollable factors, we can list commute time to the worksite and job availability in a demographic area.

According to Hinze (1985), job satisfaction has a strong effect on absenteeism. He notes that absenteeism rates are lowest when workers feel they are important members of a crew. In addition, he finds that workers with jobs that made them think have lower absenteeism rates (0.42%) than workers with jobs that required little or no thinking (3.11%). Hinze (1985) also notes the effect of management criticism. The absenteeism rate for workers who are criticized for missing work is 0.25%, compared with 1.79% for workers who are not criticized.

Job satisfaction also emerges as an important factor in turnover. Schnake and Dumler (2000) surveyed a medium-size construction company. They note that job satisfaction has the strongest relationship to turnover.

According to a Business Roundtable report (1982), workers cite the following reasons for missing work and for quitting:

- Relationship between workers and bosses
- Overtime available elsewhere
- Unsafe working conditions
- Excessive rework
- Commute time to the worksite

The study notes that self-reports are probably understated because workers are either self-conscious or have poor recall. The study also compares less-experienced workers to more-experienced workers. More-experienced workers are more concerned with the quality of their work, have a lower turnover rate, and have a higher absenteeism rate.

The Business Roundtable report (1982) also included a survey on job satisfaction, which concluded the following:

- Job dissatisfaction affected absenteeism rates more than turnover rates.
- Quality of supervision and an understanding of company goals were the most important job satisfaction factors.

- Job dissatisfaction was the strongest reason for quitting in all age categories, geographic regions, and both union and open shops.
- Generally, as the size of a job increased, job satisfaction decreased.

In addition, the report reveals that only a small percentage of the workforce is responsible for absenteeism and turnover. In other words, the same workers are continually absent. During a one-year period, 67% of the workforce reported missing only 5 or fewer days, while 5% reported missing more than 40 days. The same was true for turnover. Over a two-year period, 65% of the workforce said they rarely or never considered quitting their jobs, and only 10% said they often or very often considered quitting.

High absenteeism and turnover rates can affect any workforce's production. Hinze (1985) briefly discusses the economic impact of these problems in his research. He notes that a decrease in production leads to increased cost for the contractor and, eventually, for consumers. The Business Roundtable report (1982) goes into more detail. Following are some of the economic impacts noted in that report:

- Lost revenue because of delays
- Administrative costs for recruiting, processing, and training new employees
- Lower efficiency of new or inexperienced workers resulting in delays
- Underutilization of capital investments (tools, equipment)
- Interrupted workflow resulting in delays or rework
- Assignment of experienced workers to lower-skilled work to meet deadlines
- More demand on administrative time and resources for planning and rescheduling
- Increased overtime resulting in increased costs and employee fatigue
- Lower employee morale

The report estimates that companies lose 12 work hours for each absent worker. For turnover, the study concludes that every termination (both quits and firings) results in the loss of 24 work hours. The study states that a reduction of 50% in the annual turnover rate would cut labor costs by 1.5%. The overall conclusion is that if a company could cut an absenteeism rate of 10% and an annual turnover rate of 250% in half, it would save 9% in direct labor costs.

Some of the previous research studies include suggestions for managing absenteeism and turnover. For example, Hinze (1985) suggests that pressure from supervisors could help reduce absenteeism. He notes that making employees aware of the consequences of excessive absenteeism is effective. The Business Roundtable report (1982) states that face-to-face communication is the most effective way to get workers' attention. In addition, the study lists the following as methods for making workers aware of problems caused by absenteeism and turnover and for improving morale:

- Improving communication with bulletin boards and company newsletters
- Conducting toolbox meetings (in conjunction with safety meetings)
- Setting up workers' suggestion boxes
- Training supervisors in motivational and interpersonal skills
- Creating small workgroups with as much independence as the job allows
- Making job goals known
- Keeping workers informed about where they stand in relation to personal and company goals

The report also notes that better hiring practices and union action could help reduce high turnover rates. Managers using a good screening process can weed out job hoppers or poorly qualified workers. Unions can withhold referrals from workers with high records of quits or absences. However, this technique is considered effective only when jobs are scarce.

These suggestions for reducing absenteeism and turnover sound good, but do they actually work? There is no evidence that they do. Neither of the reports cited above put these suggestions into practice. Further research is needed to determine the effectiveness of various methods for reducing absenteeism and turnover.

As noted earlier, there is little research on absenteeism and turnover specifically for the construction industry. However, other research on these topics exists. Past studies have examined these problems in a wide variety of industries. Many researchers have found that absenteeism and turnover are related to some degree. How much is not entirely known. Different factors, such as the type of measurement used, affect the relationship. Some research has suggested a positive correlation between voluntary turnover and absenteeism. However, outside factors such as the job market could affect that relationship. Therefore, for general purposes, establishing a relationship between absenteeism and turnover is probably not practical.

The two most common absenteeism measurement techniques are frequency and time lost. Frequency refers to the number of days absent. Time lost is a measure of duration. However, researchers do not always agree on the reliability of these measurement techniques. Because absenteeism can be difficult to measure, some researchers have suggested measuring attendance instead. Measuring only attendance, however, does not allow researchers to examine the different forms of absenteeism.

Another problem in accurately measuring absenteeism involves how researchers gather the information. Researchers have used both self-reports and company records. Each presents problems for accuracy. Self-reports might give more accurate reasons for why workers are absent. However, workers may have poor recall about the actual number of days missed. Company records may be more accurate about number of days missed. However, some companies assign codes to distinguish among different reasons for absences. Thus, some reasons for missing work might be miscoded. These discrepancies can make it difficult to get a precise measurement of actual absences and the reasons for them.

Researchers also use many techniques to measure turnover. One technique measures turnover as a percentage of new hires divided by total workforce. As with absenteeism measures, there are problems in accurately measuring turnover. These problems include the following:

- It is difficult to distinguish between voluntary and involuntary turnover (quits versus firings or reductions in force).
- Tenure is sometimes used as a substitute measure of turnover.
- Some turnover information is inaccurate.

Much research exists on factors affecting absenteeism and turnover. As might be expected, almost all of the factors affecting absenteeism also affect turnover. The commonly researched factors include the following:

- Job satisfaction
- Workers' personal factors
- Safety
- Organizational factors
- Management
- Job performance

Previous studies do not agree on how absenteeism and job satisfaction are related. Some researchers find no relationship between the two, while others find a highly negative relationship. In analyzing previous research on turnover, we could not help but think that researchers could apply different measures across studies for job satisfaction. In fact, one researcher used two measures of satisfaction in his study: process satisfaction and outcome satisfaction. Process satisfaction refers to an interpersonal experience. Outcome satisfaction measures a worker's view on group composition and achievement of department objectives. We believe that using different measures for turnover can maximize the reliability of turnover prediction.

The two most studied workers' personal factors are age and tenure. We found some discrepancies among the studies done on absenteeism as it relates to age and tenure. We believe that to better understand this relationship, a widespread analysis is required.

The studies on voluntary turnover found a negative relationship between age and turnover. However, some of those studies had various problems in the research work. Age alone may not prove to be an accurate predictor of voluntary turnover. Past research shows that age and turnover are not statistically related, and many studies have shown small correlation values. The general trend in all the studies was, if anything, a slightly negative relationship between age and turnover.

It is a common belief that longer tenure is associated with lower turnover. This tenure-turnover relationship was consistent among most of the research studies. However, an overall analysis of previous research found that the relationships of tenure and age produce almost no effects on voluntary turnover.

We could derive no clear conclusions regarding the relationship of turnover to either age or tenure. Studies in which the researchers made their conclusions based on examining records of previous work generally found a negative relationship between tenure and turnover. The studies that were based on an observation of ongoing work found little or no relationship between the two. There are several causes for this variation in results:

- Measurement errors can sway results one way or another.
- Researchers may misinterpret information from company records.
- Significant differences in the worker populations studied may exist. (For example, the factors that affect, say, clerical workers may or may not also affect construction workers.)

We found no research on absenteeism and safety specifically for the construction industry. In general, researchers agree that high absenteeism can result in unsafe working conditions. For example, those who must fill in for absent workers can become more easily fatigued and make dangerous mistakes. However, researchers do not entirely agree on the factors that cause accidents. Common sense dictates that whatever the reasons may be, lowering the accident rate in the workplace is a necessary goal. Managers who are able to reduce absenteeism can have a positive effect on workplace safely.

Organizational factors encompass both company administration and personal characteristics of workers. Organizational factors for both absenteeism and turnover include the following:

- Pay
- Company size
- Work crew size and interaction
- Tasks
- Responsibilities
- Worker commitment

Previous research reveals a clear relationship between absenteeism and organizational factors. For example, large work crews generally experience higher absenteeism rates than small work crews. A high level of worker commitment results in lower absenteeism rates. Even group interactions can have an effect. One researcher found that workers' absence patterns reflected that of their workgroup. If a workgroup tolerated absenteeism, workers were absent more often. Another study found that increased pay led to increased worker commitment, while repetitive tasks reduced commitment.

Studies on how organizational factors affect turnover agree that pay, promotion, and commitment have an inverse relationship to turnover. Good salaries, opportunities for promotion, and worker commitment generally result in lower turnover rates.

Although past studies give interesting suggestions for reducing absenteeism, none of the proposed methods were tested. Some of the methods used include the following:

- Offering rewards for attendance
- Administering disciplinary action
- Improving communication with workers

Some of the research stated that these methods are effective; others did not. We believe that absenteeism prevention programs require testing and further study to fully understand what, if any, effect they have. Many of the studies on managing turnover are qualitative. As we found with the absenteeism studies, suggestions for reducing turnover were interesting but untested.

Researchers have developed several models to help management better understand and manage turnover trends. Figures 6.1 and 6.2 show two of the more widely used models.

Job performance is a relatively new factor that researchers are examining for its ability to predict turnover. Researchers have consistently found an inverse relationship between job performance and turnover. Poorer-performing workers tend to quit more often than higher-performing workers.

Based on our review of the research, we believe that companies must take a proactive approach to reducing absenteeism and turnover. Of course, employees must take part of the blame for these problems. However, companies should also examine their hiring, training, communication, pay, and promotion policies. Changes to those policies may result in more effective ways to attract and retain workers with good skills and good attendance records.

As noted earlier, researchers may disagree on how to collect and interpret information. In addition, problems that can occur in any research study may skew results and conclusions. However, the consensus is that management action with regard to controllable factors can help reduce absenteeism and turnover. As one

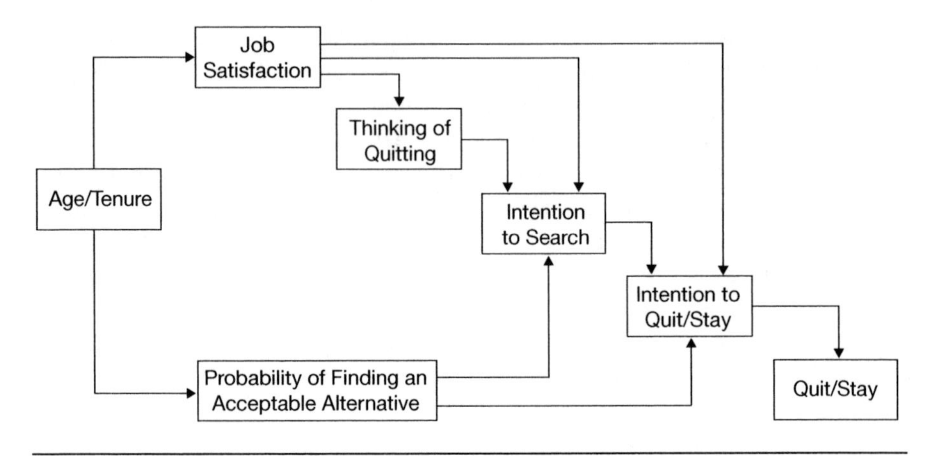

Figure 6.1 Intermediate linkage in the employee withdrawal decision (Mobley et al. 1978).

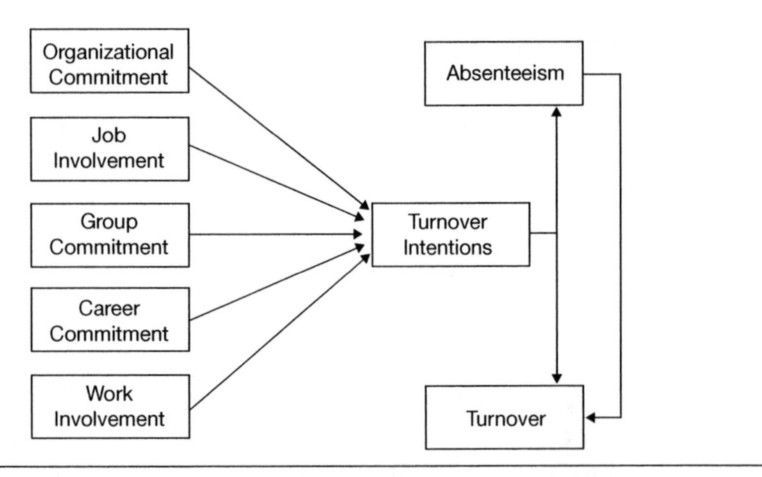

Figure 6.2 Direct model with progressive withdrawal process (Cohen 2000).

researcher stated, "An organization gets the attendance it expects, or the absenteeism it accepts."

UNDERSTANDING ABSENTEEISM AND TURNOVER

Absenteeism and turnover affect both workers and management. Crews with absent members must still get the work done and may have to double up on tasks or work with less-experienced replacements. Both situations can cause delays and may create dangerous working conditions. Managers also lose time and efficiency because they have to find replacement workers, assign alternative tasks to remaining crew members, or reschedule work. In construction, especially, delays in just one area can affect the entire project, backing up the schedules of other work crews and creating problems for everyone involved.

It is important, therefore, to learn what causes absenteeism and turnover and to learn what methods companies can use to combat these problems. To gain a better understanding of the work environment, we distributed a survey to contractors in the electrical construction industry. We designed our survey questionnaires with four objectives in mind:

1. Gain insight from both fieldworkers and managers on reasons for absenteeism and turnover
2. Identify any common agreements or disagreements between fieldworkers and managers regarding absenteeism and turnover
3. Identify any relationships between absenteeism and turnover and the parties involved
4. Identify best practices for reducing absenteeism and turnover

To get accurate information from respondents, we defined absenteeism as any missed days from work not including holidays or vacation days. We defined turnover as workers voluntarily quitting. We believed that it was important to gain insights from both management and workers. Therefore, the survey packets sent to contractors included two separate questionnaires: one for managers and one for fieldworkers. A cover letter to a member of management gave directions for properly distributing the questionnaires to the field.

We asked managers to provide information on company background, demographics, and current company policies. We also asked them to report on their specific absenteeism and turnover problems and to state why they thought their employees were absent or quit.

We asked fieldworkers to provide information on crew and management relations, work demographics, and background. Our goal was to determine the number of absences and turnovers and the reasons employees gave for missing work or quitting.

We benefited from the assistance of the University of Wisconsin-Madison Survey Center while drafting our questions. Staff at the center recommended writing the questions in an informal, conversational style. They indicated that this style would help respondents feel personally involved and more comfortable about providing information.

We mailed survey packets to randomly selected NECA contractors throughout the United States. A total of 52 managers (5.2% response rate) completed and returned questionnaires. We received 46 responses (4.6% response rate) from fieldworkers.

To help us set the scope of the analysis, we asked managers and fieldworkers questions regarding their location. The manager questionnaires came from a total of 25 different states. Of the 52 manager respondents, 37% work in the Midwest, 33% in the Southeast, 15% in the Southwest, and 13% in the Pacific Northwest. One respondent (who accounts for the remaining 2%) left that question blank.

The fieldworkers questionnaires came from a total of 26 different states. Of the 45 workers responding, 38% work in the Midwest, 24% in the Southeast, 20% in the Southwest, and 16% in the Pacific Northwest. One respondent (who accounts for the remaining 2%) works in the New England area. The geographical areas that were not well represented in either the manager or worker surveys were the New England states and the upper Midwest, particularly the Dakota region.

We asked both managers and workers to state what type of construction they typically worked on. Of the total responses, 24% work on commercial construction, 17% on industrial, 15% on maintenance, 12% on institutional, 9% on manufacturing, and 8% on residential. The remaining 15% work on power plant, wastewater treatment plant, and other types of construction.

We asked managers to report how many workers they currently employ and got 48 responses. The average number of workers employed was 64.9, with a minimum of 2 and a maximum of 500. Managers from two companies reported that their workforce fluctuates. One reported having 5 to 6 workers at any one time, and the other reported employing 15 to 30. In terms of staff size, almost all of the companies are small to midsized. The largest group of respondents (26%) included companies that employ between 11 and 20 workers. The next largest group (20%) employs fewer than 10 workers. To get a better understanding of project size, we asked workers to estimate the total number of construction workers on a typical project. We received 45 responses to this question. Most reported working on jobsites with 10 or fewer workers.

We received 45 answers to a question on commuting time. Most (93%) reported commuting 60 minutes or less, 19 commuted 30 minutes or less, and 23 commuted between 30 and 60 minutes. One worker reported that a typical job involved an overnight stay, and two reported commuting between 60 and 120 minutes.

Managers are responsible for making sure that workers are kept busy. The quality of a foreman was noted as a possible factor for absenteeism or turnover. The worker survey addressed this situation by asking whether lack of work was a problem on a typical jobsite. Most (80%) reported that it was not a problem; 20% said it was. Because lack of work is attributed to poor management, we asked workers to choose their level of agreement with a series of statements about management characteristics. Almost all (45) of the workers rated the statements on management characteristics. Generally, workers believed that their bosses did a good job of running a project. However, they apparently believed that their bosses do not consider absenteeism serious enough to warrant disciplinary action.

We asked workers to rate several statements related to coworker characteristics. Respondents generally agreed that their crew members are friendly, easy to approach, and like to help each other do a better job. Respondents were neutral on the following:

- Socializing after work
- Being asked for their input on problems in the field
- Feeling rushed on the job
- Taking a stance as a crew against a coworker's absenteeism habits

Every worker who responded to the survey was male, with an average age of 41.34 years (standard deviation of 9.02 years). The median age of the respondents was 41.5 years. The youngest respondent was 22.5 years old; the oldest was 60. Because experience is highly correlated with age, we also analyzed respondents' years of experience for our study. The majority of the respondents had between 10 and 30 years of experience. The largest group (33%) had 20 to 30 years of experience. Most of the workers responding were journeymen (28%), foremen (43%), or supervisors (22%).

Qualitative Survey Results on Absenteeism Characteristics

In this section, we focus on the answers we obtained from managers and workers regarding survey questions on absenteeism. We analyzed several important relationships found between the absenteeism data and the characteristics discussed above. To analyze the data, we used the following statistical tests:

- An analysis of variance (ANOVA) test, which tests for different means among two or more groups
- A two-sample t-test, which tests for significant differences of two groups by testing for differing means
- A chi-square test, which is used to find the differences among categorical data

Of the 52 respondents to the manager questionnaire, 17 (33%) reported that their company had a problem with absenteeism. Also, 90% of all respondents said their on-site productivity was affected more by absenteeism than by managerial productivity.

Of the 46 respondents to the electrician questionnaire, 32 (70%) reported that they were never absent in the month prior to completing the questionnaire. Of the

14 (30%) respondents who reported being absent in that month, the days absent were as follows:

- Seven (15%) missed one day
- Five (11%) missed two days
- One (2%) missed three to four days
- One (2%) missed 5–6 days

To illustrate time lost per month, let's say one month includes 23 workdays. If the 46 respondents were part of one crew, they would account for 1,058 total workdays (23 × 46). Our data show that on average, this crew would lose 26 workdays, or 2.5% of the total workdays lost for that month. This percentage is considerably lower than the 20% reported in an earlier research study. However, contractors we met with said that absenteeism normally is less than 5%. Workers also reported the number of times they had been absent during the year before our study. Thirteen (28%) said they were at work every day. Twenty-eight (61%) reported missing one to five days. Five (11%) reported missing between 5 and 10 days. To test the relationship between monthly and yearly absenteeism, we ran a correlation test. The monthly and yearly absenteeism data had a significant positive correlation. This means that a worker who misses a lot of work in a month is likely to miss a lot of work during the rest of the year.

We believe that the yearly report is, overall, a better measure of attendance than the monthly report. A worker may have been absent shortly before we administered the questionnaire. Therefore, looking only at the monthly data could make that worker's attendance record seem poorer than it actually is. On the other hand, it is difficult for workers to remember exactly how many days they missed over the course of a year. Therefore, in terms of actual days missed, the monthly absenteeism information is more accurate than the yearly information.

We asked workers to report on injuries throughout their career that caused them to miss work. Slightly more than half (52%) of the workers reported they had had a work-related injury sometime during their career that caused them to miss work. Of the workers who were not absent during the previous month, 44% reported having had injuries that caused them to miss work. Of the workers who were absent during the previous month, 71% reported having had such injuries. When we looked at the yearly numbers, we found that 50% of workers who were not absent had injuries, compared with 53% of those who were absent. We originally thought we would find positive correlations between the number of lost-time injuries and both the monthly and yearly data. However, we found a significant correlation only for the monthly data.

We used 24 responses to chart the time lost due to injuries. Although the average time lost due to injury was fewer than three days, 17% of the injuries resulted in more than a month of lost time. Workers who had been absent for more than a month because of injuries reported their total time lost as 8 weeks, 10 weeks, 3 months, or 9 months.

We also compared length of time spent on one jobsite to the monthly and yearly absenteeism numbers. Although we found no significant relationships, we did notice something that could produce significant results in future studies. Of the 13 workers who reported working on one jobsite for three to six months, almost all (12) reported being absent some amount of time during the year.

We found no significant relationships between monthly absenteeism data and management characteristics. However, the yearly data produced significant relationships. The mean rating of the group of workers with at least one absence during the year is significantly different and larger than the mean rating of the group with no absences. In other words, workers who were not absent during the year had a better opinion regarding their bosses' management characteristics than those who were absent.

To analyze data on coworker characteristics, we divided yearly absenteeism data into two groups: workers who had not been absent during the previous year and those who had been absent. We found only one statement with a significant difference in means between the two groups. That statement reads: "The attitude taken by the crew regarding absenteeism was that it was okay." We hypothesized that workers who had been absent would agree with this statement and therefore rate this question lower than those who had not been absent. A two-sample t-test (significant at a 95% confidence level) proved that our hypothesis was correct. Workers who had been absent rated this statement at roughly 4.24 (they neither agreed nor disagreed with it). Workers who had not been absent rated it at approximately 5.42 (they disagreed somewhat with the statement).

We found no significant relationships between monthly absenteeism data and age. However, we did find significance between age and yearly absenteeism data (significant at a 95% confidence level). An ANOVA test revealed the mean age for absences in a year. We arranged the data in three categories as follows:

- No absences, mean age = 46.3
- Between 1 and 5 days absent, mean age = 40.1
- Between 5 and 10 days absent, mean age = 36.5

Because the sample size was small, we moved the data from these three categories into two: no absences and absences. We then tested the new data to determine whether the mean difference in age was significant. We arranged the new data in two categories as follows:

- No absences, mean age = 46.3
- Absences, mean age = 39.6

The new data showed a significant difference in age. Our two-sample t-test was significant at a 99% confidence level. We noted that the mean age for absenteeism is the approximate age at which most people are raising children.

We compared experience (which correlates highly with age) with absences per month and per year. There were no significant relationships between experience and days missed per month. To test the relationship between experience and days missed per year, we performed a chi-square test. The test showed a significant difference between two absence categories (no absences and one to five days absent) and two experience categories (more than 15 years' experience and less than 15 years' experience). During the year, 94% of the less-experienced respondents were absent, while only 54% of the more-experienced respondents were absent. This difference was significant at 99% confidence level. We then combined the yearly absenteeism data into two categories: no absences and absences. During one year, 94% of the less-experienced electricians were absent, and 62% of the more-experienced electricians were not (significant at a 95% confidence level). The results are consistent with those regarding age. Therefore, younger, less-experienced workers were absent more often than older, more-experienced workers.

When we compared absenteeism data with the worker's position in the company (journeyman, foreman, or supervisor), we found one significant relationship. In terms of monthly absences, journeymen were more likely to be absent than supervisors. However, there was no difference in yearly absenteeism among journeymen, foremen, and supervisors.

Reasons for Worker Absenteeism as Reported by Managers

We asked managers to rate a list of reasons for workers missing work. Respondents rated the reasons on a scale of 1 to 10, with 1 being a weak factor and 10 being a strong factor for absenteeism. We have summarized the managers' reasons for absenteeism in Table 6.1. The table includes the average level of rating and standard deviation for each factor. The top five reasons for worker absenteeism reported by managers are as follows:

1. Personal and family illness
2. Simply did not feel like working
3. Doctor/dental appointments
4. Drugs or alcohol
5. Lack of responsibility

We noted an interesting variance in the injury factor. Our research indicated that, from the worker's viewpoint, injury plays a significant role in absenteeism. In our study, managers rated injury as the eighth strongest factor. However, the standard deviation of 3.33 is higher for injury than for any other absenteeism factor.

Table 6.1 Worker Absenteeism Factors as Reported by Managers

Absenteeism Factor	Average Response	Standard Deviation
Personal and family illness	6.37	2.88
Simply did not feel like working	5.75	3.01
Doctor/dental appointments	5.61	2.48
Drugs or alcohol	4.76	3.04
Lack of responsibility	4.33	3.11
Bad weather	3.82	2.64
Boredom with job	3.47	1.99
Injury	3.39	3.33
Bad relations with boss/coworker	3.14	2.16
Travel distance	3.08	2.12
Time pressure	2.47	2.11
Excessive rework	2.20	1.88
Too much overtime	2.18	1.81
Wage rate	2.14	2.12
Excessive change orders	2.10	1.57
Inadequate tools and equipment	1.96	1.55
Poor craft supervision	1.92	1.48
Poor overall management	1.88	1.64
Unsafe working conditions	1.82	1.85
Inadequate safety plan	1.59	1.46

This means that the injury factor could actually be stronger or weaker than its current average. The injury factor was rated 1 (weak) by 25 respondents and 10 (strong) by 6 respondents. Therefore, injury as a factor in absenteeism is either insignificant or significant. We believe that the different ways companies prioritize safety accounts for this variance in ratings.

We performed a two-sample t-test to determine whether there were any differences in reasons for missing work between companies with absenteeism problems and companies without problems. We compared the two groups on the following factors for absenteeism:

- Lack of responsibility
- Simply did not feel like working
- Drugs or alcohol

Table 6.2 Statistically Significant Differences between Companies with and without Absenteeism Problems for Absenteeism Factors

Absenteeism Factor	Average Response for Companies with Absenteeism Problems	Average Response for Companies without Absenteeism Problems	Confidence Level of Significance (%)
Unsafe working conditions	2.24	1.00	99
Inadequate safety plan	1.88	1.00	99
Injury	4.35	1.47	99
Personal and family illness	6.94	5.00	95
Poor overall management	2.12	1.41	90
Excessive rework	2.44	1.71	90
Bad weather	4.26	2.94	90
Simply did not feel like working	4.85	7.53	99
Drugs or alcohol	4.15	6.00	95

We found that the mean ratings for two of these factors—"simply did not feel like working" and "drugs or alcohol"—were larger in companies without absenteeism problems. The factor "lack of responsibility" was not greatly significant. Table 6.2 shows a summary of the results for the statistically significant differences.

An ANOVA test determined that significant differences exist for absenteeism in different parts of the country. Managers from four areas of the United States rated the following three absenteeism factors:

- Unsafe working conditions
- Bad relations with boss/coworkers
- Lack of responsibility

We found significant differences among the four areas of the United States. Table 6.3 shows the values per region for those factors with statistically significant differences at a confidence level of 90%. Respondents from the Pacific Northwest, who did not report any absenteeism problems, rated two reasons higher than the other geographical areas. For the factor "lack of responsibility," we determined a significant difference on only the ratings from the Southeast and the Midwest.

To determine if company size affects absenteeism, we analyzed manager-reported absenteeism data against data on company size. We found a statistically significant correlation (significant at a 99% confidence level) between the factor "too much overtime" and both dollar value of annual volume and total annual

Table 6.3 Statistically Significant Differences among U.S. Locations for Absenteeism Factors

	Average Rating per Region			
Absenteeism Factors	**Midwest**	**Pacific NW**	**Southeast**	**Southwest**
Unsafe working conditions	1.47	3.38	1.65	1.43
Bad relations with boss/coworker	2.68	4.88	3.00	2.71
Lack of responsibility	3.16	5.00	5.71	3.43

worker hours of direct labor. The result of this analysis suggests that, in terms of volume and worker hours, larger firms produce more overtime work.

Reasons for Worker Absenteeism as Reported by Workers

We asked workers to rate a list of reasons for missing work. Respondents rated the reasons on a scale of 1 to 10, with 1 being a weak factor and 10 being a strong factor for absenteeism. We have summarized the workers' reasons for absenteeism in Table 6.4. Our sample size for this part of the questionnaire was only 26 because we instructed workers who had not been absent for the covered period to skip this section. The top five reasons for absenteeism reported by workers are as follows:

1. Personal and family illness
2. Injury
3. Doctor/dental appointments
4. Bad weather
5. Unsafe working conditions

We found two significant correlations in the data. Monthly absenteeism was significantly correlated with the factor "wage rate" (significant at a 90% confidence level). Yearly absenteeism was significantly correlated with the factor "simply did not feel like working" (significant at a 99% confidence level).

To determine whether there were any differences in average ratings of absenteeism factors between workers who had not been absent during the month and those who had, we did a two-sample t-test. Only two factors showed any significance:

- "Boredom with job" was significant at a 90% confidence level.
- "Doctor/dental appointments" was significant at a 95% confidence level.

Our analysis indicates that the absent group misses work more often for illegitimate reasons. We tested absenteeism factors with the amount of time lost to

Table 6.4 Worker Absenteeism Factors as Reported by Workers

Absenteeism Factor	Average Response	Standard Deviation
Personal and family illness	6.04	2.95
Injury	5.15	3.60
Doctor/dental appointments	4.80	2.74
Bad weather	4.35	3.05
Unsafe working conditions	3.04	2.75
Poor overall management	2.69	2.31
Inadequate safety plan	2.54	2.77
Simply did not feel like working	2.42	2.04
Poor craft supervision	2.31	1.83
Excessive rework	2.27	2.11
Bad relations with boss/coworker	2.19	1.79
Travel distance	2.00	1.57
Inadequate tools and equipment	1.88	1.77
Wage rate	1.85	1.71
Excessive change orders	1.81	1.52
Boredom with job	1.81	1.74
Too much overtime	1.77	1.45
Drugs or alcohol	1.69	2.19
Lack of responsibility	1.58	1.39
Time pressure	1.50	1.10

injuries over a career. A correlation test produced a moderately significant relationship with the factor "too much overtime." Although the significance level is not very high, the results of the test suggest that workers working overtime might become more fatigued or less careful, thus increasing the risk for injury.

We found significant correlations between absenteeism factors and size of a construction site. As might be expected, workers noted "poor craft supervision" as an important factor on larger jobsites. They also reported that working conditions got worse as the size of a job increased. They reported that there was more pressure to complete the project and an increase in drug and alcohol abuse.

We studied the relationships between absenteeism factors and management characteristics to better understand how a manager's actions affect a worker's absenteeism or excuses for missing work. For example, we found a strong relationship between the factor "poor overall management" and the statement "Your boss was understanding of workers' personal problems and tried to help them." Workers who said that "poor overall management" was a significant factor for their absenteeism also said that their bosses were less understanding of personal problems. This correlation suggests that management plays a key role in a worker's absenteeism. The factor "lack of responsibility" was also positively correlated with three management characteristics:

- Understanding workers' personal problems
- Work ethic
- Ability to match workers to assignments

Coworker relationships and reasons for absenteeism generated two significant correlations. Both correlations were significant at a 95% confidence level. Workers who were absent because of their relationship with their boss or a coworker also spent more time outside of work with coworkers. Workers who reported that their bosses asked crews to redo a lot of work were more likely to claim excessive change orders as an excuse for missing work.

When we compared age to reasons for absenteeism, we found that younger workers were more likely than older workers to list "company safety plans" and "excessive rework" as excuses for absenteeism. Older workers appeared to take more days off because of bad weather. We broke age down into two groups: younger (age 40 and younger) and older (age 41 and older) and tested for the mean differences between the two groups. There were 13 responses in each group. The younger group rated the following four absenteeism factors higher than the older group:

- Unsafe working conditions (younger group: 4.08, older group: 2.00)
- Inadequate company safety plans (younger group: 3.62, older group: 1.46)
- Excessive rework (younger group: 3.15, older group: 1.39)
- Too much overtime (younger group: 1.77, older group: 1.23)

Our test results showed that the factors "unsafe working conditions" and "inadequate company safety plans" were significant at a 90% confidence level. The factors "excessive rework" and "too much overtime" were significant at a 95% confidence level. Older workers rated the factor "bad weather" higher than younger workers. This factor accounted for the largest difference between the two groups. The older workers rated it 6.08, while the younger electricians rated it 2.62.

Comparison of Reasons for Worker Absenteeism between Managers and Workers

We compared the managers' reasons for absenteeism to the workers' reasons. Overall, the responses from the managers had a higher average rating (3.20) compared with the responses from the workers (2.64). Table 6.5 lists the nine absenteeism factors with statistically significant differences. The column titled

Table 6.5 Statistically Significant Differences in Management Responses versus Worker Responses on Absenteeism Factors

Absenteeism Factor	Average for Managers	Average for Workers	Estimated Difference	Significance Level (%)
Unsafe working conditions	1.82	2.96	-1.14	90
Injury	3.39	5.00	-1.61	90
Travel distance	3.08	1.96	1.12	99
Boredom with job	3.47	1.78	1.69	99
Bad relations with boss/coworker	3.14	2.15	0.99	95
Lack of responsibility	4.33	1.56	2.77	99
Simply did not feel like working	5.75	2.37	3.38	99
Time pressure	2.47	1.48	0.99	99
Drugs or alcohol	4.76	1.67	3.09	99

"Estimated Difference" is the difference between the means of each group. The largest differences are associated with four factors that managers have ranked higher than electricians:

- Boredom with job
- Lack of responsibility
- Simply did not feel like working
- Drugs or alcohol

None of these factors is controllable by management. A graphic representation of the average response for each factor is shown in Figure 6.3. The factors most agreed on by both workers and managers were "personal and family illness," "bad weather," and "doctor/dental appointments." These three factors may, in fact, be the top three reasons for absenteeism in the electrical construction industry.

Methods for Reducing Absenteeism as Reported by Workers:

We gave workers a list of nine suggestions for reducing absenteeism and asked them to choose as many suggestions as they wished. We included a space for them to suggest other methods but got only one suggestion: a monetary reward program for perfect attendance. Most of the workers chose incentive programs (69%) and a four-day 40-hour workweek (49%) as the top two ways to reduce absenteeism. Figure 6.4 shows all answers given.

We then asked workers to choose only one of the nine methods that would most help absenteeism. Because 4 of the 46 respondents did not answer this question, we used 42 answers for our analysis. Once again, most of the workers chose

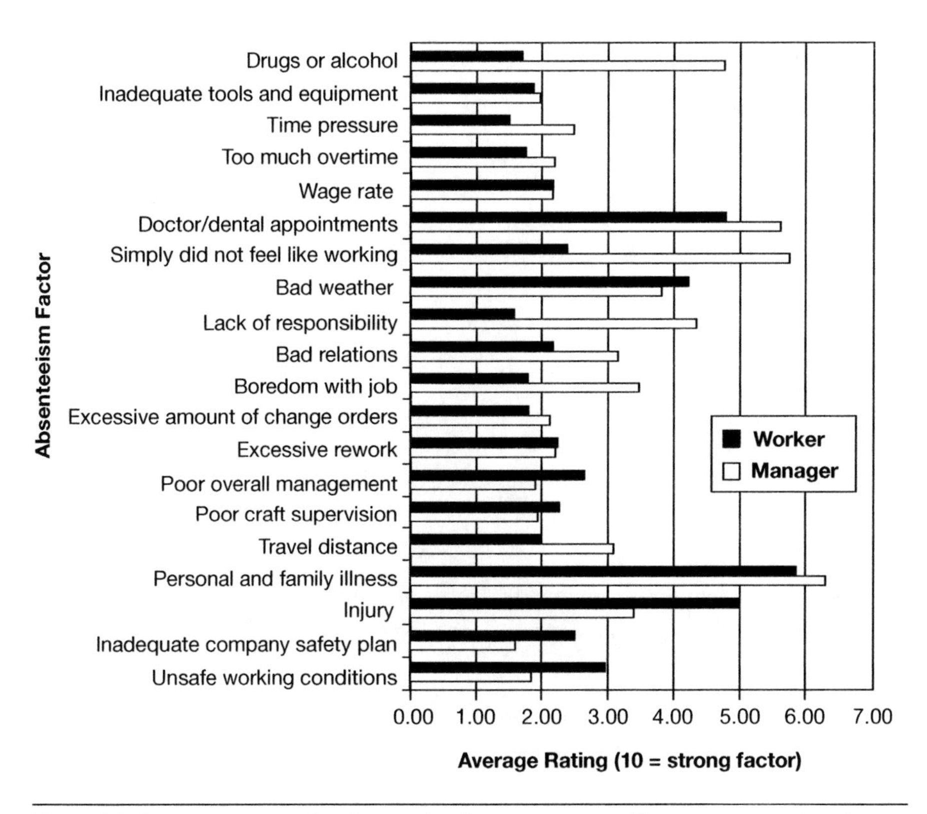

Figure 6.3 Average response for absenteeism factors as reported by managers and workers.

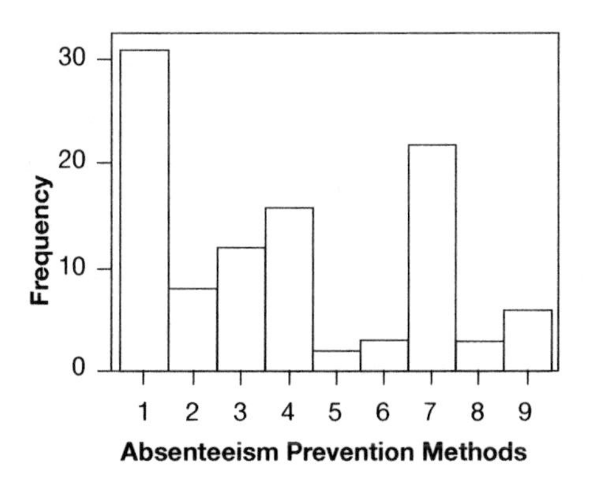

1 = Incentive programs
2 = Safer sites
3 = Better management
4 = More recognition
5 = Less overtime
6 = More overtime
7 = 4-day, 40-hour workweek
8 = 5-day, 40-hour workweek
9 = Better training

Figure 6.4 Methods for reducing absenteeism as reported by workers.

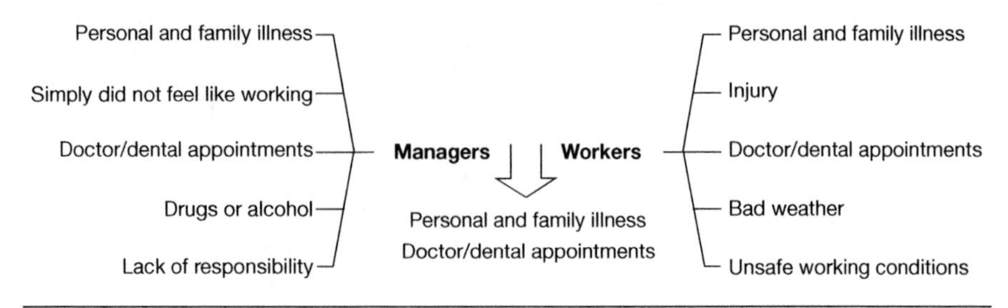

Figure 6.5 Comparison of manager's and worker's top reasons for absenteeism.

incentive programs (45%) and a four-day 40-hour workweek (31%). Four respondents chose more recognition (10%), three chose better training (7%), two chose better management (5%), and one chose less overtime (2%).

Survey Response Summary for Absenteeism

Based on the information provided by managers and workers, we can make several conclusions regarding absenteeism. We have summarized the results of our survey as follows:

1. Of the top five reasons given for absenteeism by managers and workers, there was agreement on two factors: "personal and family illness" and "doctor/dental appointments" (Figure 6.5).
2. Companies with an absenteeism problem rate three factors that relate to a worker's attitude ("lack of responsibility," "simply did not feel like working," and "drugs or alcohol") higher than companies without a problem. Management cannot control these factors.
3. Southeastern managers rated "lack of responsibility" as a factor for absenteeism significantly higher than Midwestern managers.
4. If we consider the 46 responding workers as one crew with an average of 23 working days per month, the self-reported monthly absenteeism from this survey would result in a minimum of 2.5% of total workdays lost.
5. A majority (72%) of the respondents missed at least one day of work in the previous year.
6. Workers named incentive programs and a four-day 40-hour workweek as the preferred ways to reduce absenteeism.
7. Younger, less-experienced workers were absent more than older, more-experienced workers. Journeymen were more likely to be absent than foremen or supervisors.

8. Bad weather was more of a problem for older workers (age 41+) than for younger workers.
9. Workers who were more likely to be absent were also more negative toward management.
10. Workers who were not absent during the previous year reported that the crew they worked with most believed that coworker absenteeism was not okay.
11. Workers who were absent during the previous month were more likely to have lost time during their career as a result of injury. A majority (71%) of workers who were absent in the previous month were injured at least one time during their career, compared to 44% of workers who were not absent. Also, workers rated the factors "injury" and "unsafe working conditions" statistically higher than managers.
12. Injuries showed the highest standard deviation of all the reasons for absenteeism among both managers and workers. This suggests that injuries often play a major role in absenteeism.
13. Managers reported that company size (total dollar value of annual volume and total annual worker hours of direct labor) shows a significant and positive correlation with the factor "too much overtime."
14. As the length of time spent at the same jobsite increased, workers felt that working conditions and poor craft supervision became greater factors for absenteeism.

Qualitative Survey Results on Turnover Characteristics

In this section, we will discuss the questions we asked managers and workers about turnover and various work-related characteristics. We used two-sample t-tests and chi-square tests to analyze several important relationships revealed by their answers. For the purpose of this survey, we defined turnover as workers voluntarily quitting a job.

We asked managers to state whether their company had a turnover problem. There were 52 questionnaires, and 51 managers responded to this question. Of those respondents, 10 (20%) reported that their companies had a turnover problem. In addition, 30 (60%) of the respondents reported that when workers quit their jobs, on-site productivity is affected more than managerial productivity.

In terms of number of workers employed, a two-sample t-test showed a significant difference (significant at a 90% confidence level) in the averages between companies that had problems with turnover and those that did not. On average, companies without a turnover problem employed 71 workers, and companies with a turnover problem employed 32.7 workers.

We originally believed that companies employing more workers would have a greater turnover problem. Our test results show the opposite to be true. However, this analysis does not prove that smaller companies experience greater turnover. If we remove the statistical outliers in these data, we note no difference between companies with and without a turnover problem. Therefore, the test might merely indicate that any amount of turnover has a greater effect on smaller companies. A statistical outlier is a value that is very different from almost all other values in an analysis. The presence of outliers can distort the true nature of the distribution of data.

We asked workers to report the number of companies they worked for in the past 10 years. Of the 46 responses received, 35 (76%) worked for three or fewer companies. These respondents have less experience than those who worked for more than three companies. It is possible that these less-experienced workers had a high degree of company loyalty, which would result in lower turnover. Of those who had worked for more than three companies, most had worked for four to seven companies. One respondent reported working for 18 companies, one for between 10 and 15 companies, and one for between 15 and 20 companies.

We also asked respondents to report the number of times they had quit a job. Half said they had never quit a job, and half said that they had. We tested the relationship between turnover and job commute using tabulated statistics and a chi-square test. Turnover was significantly greater for workers who commuted more than 30 minutes. Of those who quit at least one job, we found that 30% of those with less than a 30-minute commute quit, while 62% of those with more than a 30-minute commute quit.

We also looked at the relationship between the number of companies worked for and the statement "the crew believes that absenteeism is okay." The comparison here was between those who worked for between one and two companies and those who worked for three or more companies. We theorized that those who worked for two or fewer companies were more likely to discourage absenteeism. A two-sample t-test (significant at a 99% confidence level) proved our theory true.

The responses of journeymen, foremen, and supervisors were also compared against each other for number of quits in a career and number of companies worked for in the last 10 years. We theorized that journeymen would work for more companies than foremen or supervisors would. In fact, our analysis showed that 93% of the journeymen surveyed worked for more than one company, compared with only 61% of foremen and supervisors.

Table 6.6 Turnover Factors as Reported by Managers

Turnover Factor	Average Response	Standard Deviation
Work closer to home	6.24	2.74
Overtime available elsewhere	6.18	3.22
Indoor work versus outdoor work	4.46	2.74
Relationship with other workers	4.18	2.72
Relationship with boss	4.16	2.67
Clean jobs versus dirty jobs	3.00	2.19
Better benefits package elsewhere	2.96	2.93
Not enough recognition	2.92	2.29
Free parking versus pay for parking	2.82	2.50
Excessive surveillance (by supervisor)	2.61	2.00
Poor planning on the job	2.53	1.75
Poor craft supervision	2.35	1.81
Inadequate tools and equipment	2.25	1.84
Safer sites elsewhere	1.84	1.47

Reasons for Worker Turnover as Reported by Managers:

We asked managers to choose from a list of reasons for quitting a job. Respondents rated the reasons on a scale of 1 to 10, with 1 being a weak factor and 10 being a strong factor for turnover. We have summarized the managers' reasons for turnover in Table 6.6. Each factor received 51 responses. Managers reported the top five reasons for worker turnover as follows:

1. Work closer to home
2. Overtime available elsewhere
3. Indoor work versus outdoor work
4. Relationship with other workers
5. Relationship with boss

We also provided respondents with the option to include other reasons. The "other" category included such items as the following:

- Easier to hire workers on a bigger job
- Day work versus night work
- Lack of enough work
- Wages
- Boredom
- Union issues

Table 6.7 Turnover Factors as Reported by Workers

Turnover Factor	Average Response	Standard Deviation
Work closer to home	5.62	3.79
Overtime available elsewhere	5.19	3.52
Relationship with boss	5.14	3.50
Excessive surveillance (by supervisor)	4.48	3.27
Better benefits package elsewhere	4.24	3.69
Relationship with other workers	3.43	2.73
Clean jobs versus dirty jobs	3.33	2.87
Poor planning on the job	3.33	3.10
Not enough recognition	3.19	2.50
Poor craft supervision	3.14	2.59
Inadequate tools and equipment	3.00	2.57
Indoor work versus outdoor work	2.86	2.78
Free parking versus pay for parking	2.38	2.54
Safer sites elsewhere	2.29	2.05

We found significant correlations between reasons for turnover and percentage of indoor work on several turnover factors. Those factors are as follows (the percentages in parentheses show the level of significance):

- Excessive surveillance (99%)
- Relationship with other workers (99%)
- Clean jobs versus dirty jobs (95%)
- Poor planning on the job (95%)
- Inadequate tools and equipment (99%)
- Poor craft supervision (99%)
- Safer sites elsewhere (99%)
- Better benefits package elsewhere (90%)

We found high correlations for the factors "inadequate tools and equipment" and "safer sites elsewhere." These correlations indicate that companies with a higher percentage of inside work do not consider these factors as strong reasons for turnover.

Reasons for Worker Turnover as Reported by Workers

We instructed workers who had never quit a job to skip the questions dealing with reasons for quitting. Therefore, we received only 21 responses. Respondents rated the reasons on a scale of 1 to 10, with 1 being a weak factor and 10 being a strong factor for turnover. We have their responses in Table 6.7. Workers reported the top five reasons for worker turnover as the following:

1. Work closer to home
2. Overtime available elsewhere
3. Relationship with boss
4. Excessive surveillance
5. Better benefits package elsewhere

The questionnaire included an "other" category for workers to provide any reasons not listed. Responses included these:

- Wanted to become self-employed
- Employer violated union contract
- Slow job (not enough work)

According to the workers, as the size of the job increased, relationships with their bosses became weaker and worksites became less safe. A comparison of job size with the turnover factor "relationship with boss" was significant at a 90% confidence level. A comparison of job size with the turnover factor "safer sites elsewhere" was significant at a 95% confidence level.

We also compared turnover factors with length of time spent at one jobsite. As time at one jobsite increased, workers were more likely to rate "better benefits package elsewhere" as an important factor in deciding whether to quit. We performed a two-sample t-test by placing the length of time data into two categories: less than three months at a site and more than three months. Those who had worked at one site for more than three months rated the factor "better benefits package elsewhere" at 5.58. Those who had worked at one site for less than three months rated the factor at only 2.44. Therefore, a competitive benefits package is a strong factor in reducing turnover.

We also found that as commute time increased, the factors "poor planning on the job," "safer sites elsewhere," and "not enough recognition" became less important reasons for quitting, while the factor "better benefits package elsewhere" became more important.

To understand how management actions affect quitting or reasons for quitting, we correlated turnover factors with management characteristics. In general, the correlations give a strong indication that management actions play a key role in a worker's decision to quit a job.

We noted that management does not always consider the relationship between praise for good work and the turnover factor "relationship with coworkers." Our analysis indicates that, interestingly enough, praise is a strong factor for turnover. We believe that the way management gives praise to workers explains this circumstance. When managers praise one worker in front of the entire crew

Table 6.8 Statistically Significant Differences in Management Responses versus Worker Responses on Turnover Factors

Absenteeism Factor	Average for Managers	Average for Workers	Estimated Difference	Significance Level (%)
Excessive suveillance (by supervisor)	2.61	4.48	-1.87	95
Indoor work versus outdoor work	4.39	2.86	1.53	95

instead of privately, the crew may resent that worker. Thus, giving praise publicly might result in some resentful workers leaving. Of course, other factors come into play, but managers might want to consider crew relationships when praising individual workers.

The turnover factor "overtime available elsewhere" significantly correlated with several management characteristics. Workers rated this factor more highly as an excuse for quitting when the following management characteristics were present:

- Less open to suggestions from workers
- Unclear instructions to workers or poorly planned work
- Less knowledge of what was expected of workers

We found a few significant correlations between coworker relationships and reasons for turnover. Workers who felt that coworkers were not friendly or easy to approach reported the following:

- Their relationship with the boss as a strong factor for quitting.
- Lack of adequate tools and equipment as a strong factor for quitting. (The implication here is that workers who do not get along will not readily share tools and equipment.)
- Wanting to work closer to home as a significant reason for quitting a job. (Having friends on the jobsite might make a longer commute more bearable.)

Analysis of coworker relationships also produced the following information:

- Workers who socialize with coworkers also felt that "overtime being available elsewhere" was an important factor for quitting a job.
- Workers who socialize with coworkers do not need recognition from their superiors as much as those who do not socialize.
- Workers who agreed with the statement "You had to redo a lot of work because of poor management" reported that poor planning was a strong reason for quitting.

When we analyzed age and turnover, we found that older workers were more inclined to rate the factor "free parking versus pay for parking" as a reason for turnover. Our tests also found that older workers rated the factor "not enough

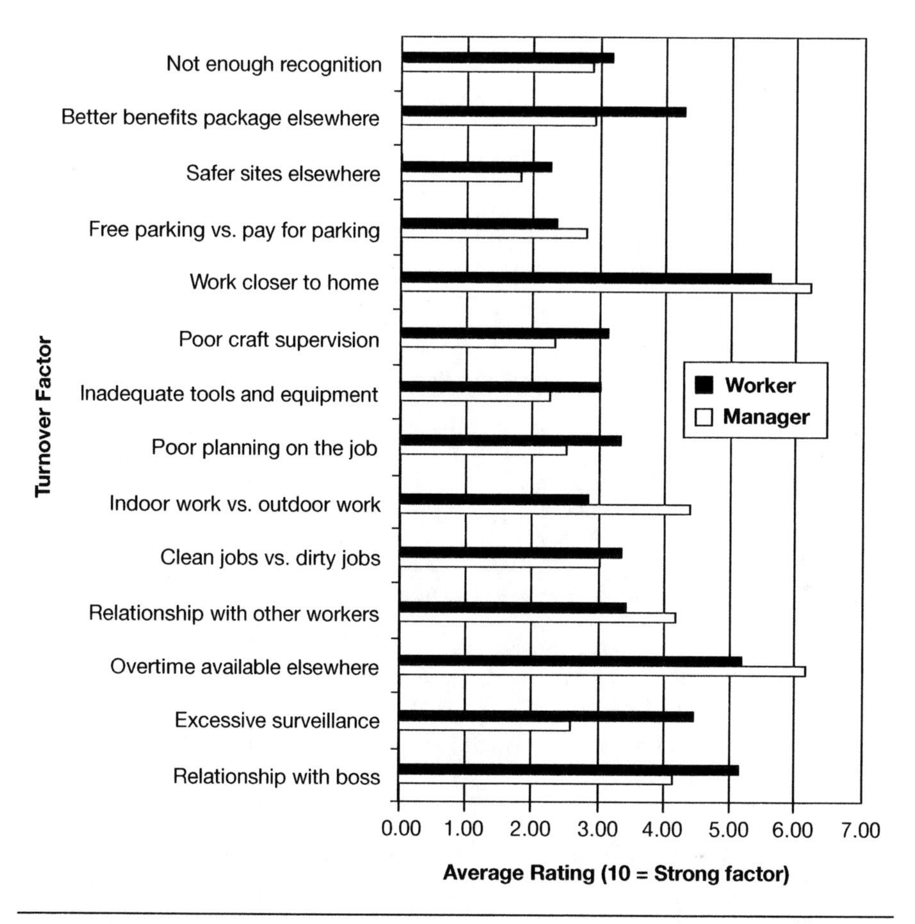

Figure 6.6 Average response for turnover factors as reported by managers and workers.

recognition" as a reason for quitting less often than younger workers. The mean rating for this factor among workers younger than 45 was 4.20, while the mean rating among workers older than 45 was 2.27.

Comparison of Reasons for Worker Turnover between Managers and Workers

We compared managers' reasons for turnover with those given by workers. Our analysis produced significant differences between the two sets of reasons on only two factors: "excessive surveillance" and "indoor work versus outdoor work" as shown in Table 6.8. This analysis indicates that electricians may place a higher value on their relationship with management than they do on whether they work indoors or outdoors. Figure 6.6 shows a comparison of the average response for

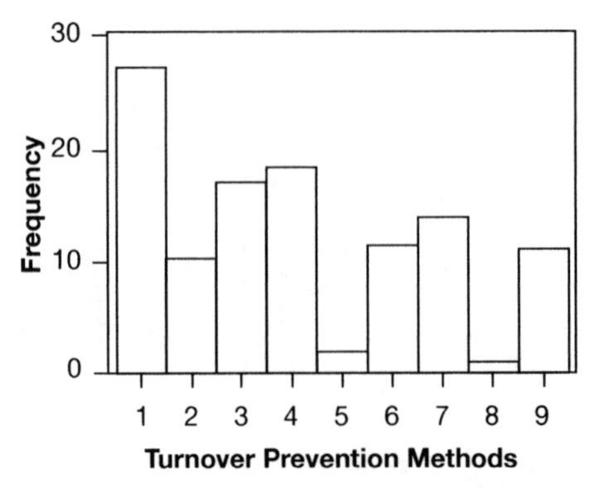

1 = Incentive programs
2 = Safer sites
3 = Better management
4 = More recognition
5 = Less overtime
6 = More overtime
7 = 4-day, 40-hour workweek
8 = 5-day, 40-hour workweek
9 = Better training

Figure 6.7 Methods for reducing turnover as reported by workers.

each turnover factor given by managers and workers. Both groups rated the factors "overtime available elsewhere" and "work closer to home" as the top two reasons for turnover. Factors with a rating higher than 3.00 indicate close agreement between both groups.

Methods for Reducing Turnover as Reported by Workers

We gave workers a list of nine suggestions for reducing turnover and asked them to circle as many as they wished. Out of the 46 respondents, 45 chose at least one suggestion as shown in Figure 6.7. Most of the respondents (60%) chose incentive programs. In the space on the questionnaire marked "other," respondents provided the following additional suggestions:

- More money
- Treat people as if they are important
- Fewer jobs to go to
- Steadier work

We next asked workers to choose only one of the nine methods that would most reduce turnover. Once again, most of the workers chose incentive programs (34%), followed by more recognition (34%), better management (11%), and four-day 40-hour workweek (9%).

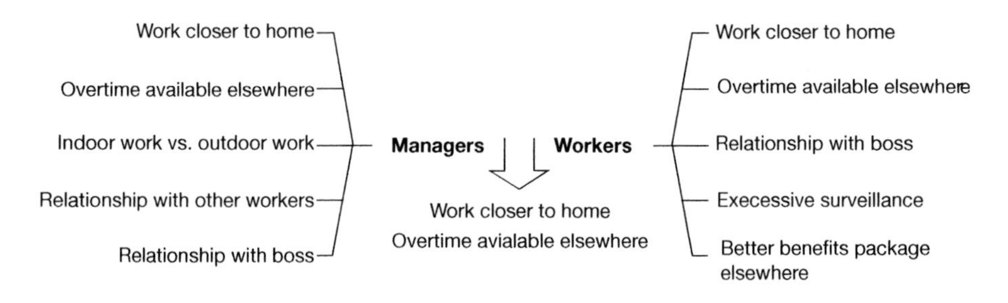

Figure 6.8 Comparison of manager's and worker's top reasons for turnover.

Survey Response Summary for Turnover

Based on the information provided by managers and workers, we can make several conclusions regarding turnover. We have summarized the results of our survey as follows:

1. Of the top five reasons given for turnover by managers and workers, there was agreement on two factors: "work closer to home" and "overtime available elsewhere," as shown in Figure 6.8.
2. Of the responding companies, 20% had turnover problems.
3. Regarding the number of companies worked for over the past 10 years, 76% of the 46 responding workers worked for three or fewer companies.
4. Half of the responding workers said they quit at least one job during their career.
5. The two most preferred methods for reducing turnover were incentive programs (first choice) and more recognition.
6. Turnover was significantly higher for workers who commuted more than 30 minutes to the jobsite.
7. In terms of job position and number of companies worked for, 93% of the journeymen worked for more than one company, and 61% of the foremen and supervisors worked for more than one company.
8. The factors "excessive surveillance" and "work closer to home" were very highly correlated with number of quits in a career and number of companies worked for in the past 10 years.
9. Among workers who had long commutes, "poor planning on the job," "safer sites elsewhere," and "not enough recognition" were less important as reasons for turnover, and "better benefits package elsewhere" was more important.
10. As length of time at one jobsite increased, the company benefits package became a more important factor in turnover.

11. As the size of the job increased, two factors, "relationship with the boss" and "safer worksites," became more important factors in turnover.
12. Older workers rated "not enough recognition" as less of a turnover factor than younger workers did.
13. Workers who reported quitting a job or who had worked for more than two companies in the last 10 years reacted more negatively toward management.
14. Workers who did not get along with their crew members cited "inadequate tools and equipment" as an important reason for turnover. They also wanted a shorter commute more often than workers who did get along with their crew.
15. Relationships with coworkers suffered when management praised good work. (Coworkers may resent a worker who is publicly praised by management.)

QUANTITATIVE ANALYSIS

In the previous sections we illustrated and discussed the work-related factors that cause absenteeism and turnover. However, we wanted not only to focus on what causes these problems but also to determine their effect on labor productivity. Our research has produced explanatory rather than predictive models. Predictive models generally require a large amount of data, and researchers use them to predict future trends in specific areas. Researchers create explanatory models using relatively small amounts of data and use them to explain variances in that data. Although explaining these variances was an important part of our study, we also identified general, overall trends, which are equally important to contractors.

The qualitative part of our research consisted of two sections: a microanalysis and a macroanalysis. A microanalysis examines a specific, ongoing project or projects. A macroanalysis examines an entire project or projects after completion. We measured both types of projects by input versus output.

We must point out that our research samples were small, and the data we got for absenteeism and turnover were to some extent limited. Therefore, those wishing to use our results on larger samples or to examine companies with higher rates of absenteeism or turnover should proceed with caution. Such a study will probably require building a larger database.

Microanalysis

To get information for the microanalysis, we asked electrical construction companies to collect approximately four to five weeks' worth of data on an ongoing project. Companies were asked to track the following information:

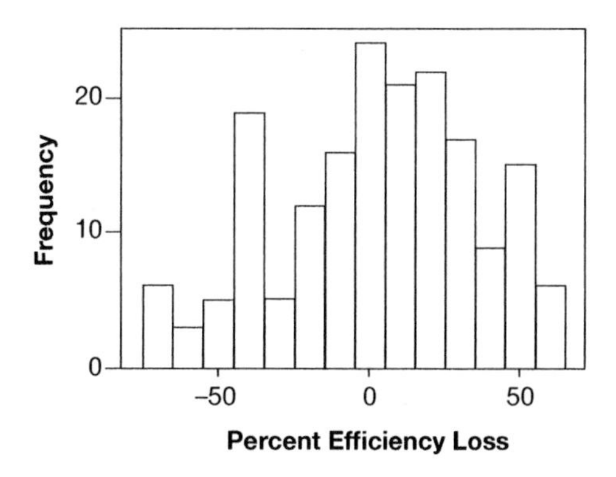

Figure 6.9 Range of values for percent efficiency (productivity) loss.

- Type of construction activities
- Budgeted production for each activity
- Number of units installed that day for each activity
- Number of crew members assigned to each activity
- Number of replacement workers required for each activity
- Number of absentees
- Total number of hours worked each day

We collected data for the microanalysis from eight different electrical contractors on eight projects. However, because three of the data collection forms were incomplete, we analyzed only five projects. Data characteristics define the scope of the research. The complete data collection forms we received covered projects in Connecticut, Florida, Michigan, New York, and Wisconsin. The electrical construction activities listed by the responding contractors include the following:

- Installing light fixtures, switch gear boxes, junction boxes, conduit, lay-in fixtures, and apartment trim
- Pulling cable
- Setting pole bases for parking lot lights
- Roughing in slabs and wiring for alarms and telecommunications (tele/data) equipment

We defined loss of productivity (or efficiency) as the ratio of the difference in actual daily production and estimated production to the actual daily production. We defined production as worker hours per unit.

The average productivity loss was 2.17%, with a minimum of –72% and a maximum of 63%. We found a wide distribution of values for percent productivity loss around a median of 5.13% as shown in Figure 6.9. Note that these values are not necessarily losses but differences from the estimate. Previous research indicates that the typical range of accuracy for electrical construction estimates is plus-or-minus 5%.

To determine the effect of absenteeism on productivity, we separated the results of calculations on lost productivity into two groups: days without absences and days with absences. There were 141 data points for days without absences and 39 data points for days with absences. The average productivity loss for days without absences was –0.9% (negative percentage indicates a higher productivity than estimated). The average productivity loss for days with absences was 13.3%. This difference was found to be statistically significant at a 98% confidence level.

Macroanalysis

For our macroanalysis, we asked the electrical companies to collect data on a completed project. The questionnaire we sent included four sections to gather information on the following:

- Company background
- A past project with higher than normal absenteeism or turnover
- Workforce
- Productivity

Our target group included 500 randomly selected NECA contractors. Because our earlier research showed that absenteeism is higher in the South, most of the contractors on this list are located in that part of the country. Our first mailing produced a low response, so we sent approximately 175 more surveys to companies in Texas, Pennsylvania, Georgia, New Jersey, and Michigan. It took roughly six months to collect the data for the macroanalysis. We got responses from 40 contractors, who provided data on 54 projects. After reviewing the questionnaires, we excluded any that were incomplete, which left 31 projects from 26 contractors for the analysis.

As noted earlier, data characteristics define the scope of the research. Contractor data characteristics show the range of company size, the types of construction in which their companies work, and the types of electrical work performed. These characteristics also define the scope for the final models. The final results are thus applicable only to projects with similar characteristics. Contractors who responded to this questionnaire are located in 14 states. The highest concentration is in the South and the area around the Great Lakes. We asked contractors to provide their annual volume for work they perform in each of the following types of construction:

- Commercial
- Industrial
- Institutional
- Manufacturing

- Power plant
- Wastewater treatment plant
- Residential
- Maintenance
- Other

Commercial work (37.19%) and industrial work (19.35%) account for the largest part of the respondents' average annual volume. We also asked contractors to report how much work they perform (in annual volume) in the following categories:

- Inside electrical work
- Residential
- Telecommunications (tele/data)
- Other

Most of the companies performed inside electrical work (73.85% of annual volume). To set the scope in terms of company size, we asked contractors to report their total annual volume. The average company's annual volume was $52 million. The data would seem to indicate that our study encompasses a wide range of contractors, both large and small. However, most of the respondents are small contractors, with half of them averaging less than $5.5 million of annual volume.

Contractors completed all of the construction projects included in our macroanalysis after the year 2000. The completed projects were located in several states. The greatest concentration of projects was in the South and the Great Lakes area. Most of the projects (11) were institutional, followed by 9 commercial projects, 5 power plant projects, 4 industrial projects, 1 manufacturing project, and 1 casino project.

To further define the type of work performed, we also placed the completed projects into three other categories: new construction, renovations, and additions. Most of the projects (17) were new construction, followed by 8 renovations, and 6 additions.

We measured project size in contract hours. Most of the projects (68%) had fewer than 40,000 total contract hours. We recommend caution in using the results of our research when analyzing projects with more than 200,000 or fewer than 2,000 contract hours. Our study received limited data for projects of those sizes.

The shortest project lasted 10 weeks, and the longest project lasted 156 weeks, or almost 3 years. The average project duration was 51.26 weeks. Again, we advise caution when using these results to analyze projects lasting fewer than 10 weeks or more than 90 weeks.

Contractors reported absenteeism as a percentage of the number of workers absent to the number of workers employed. We broke down project absenteeism into three percentage categories: 0–5%, 6–10%, and more than 10%.

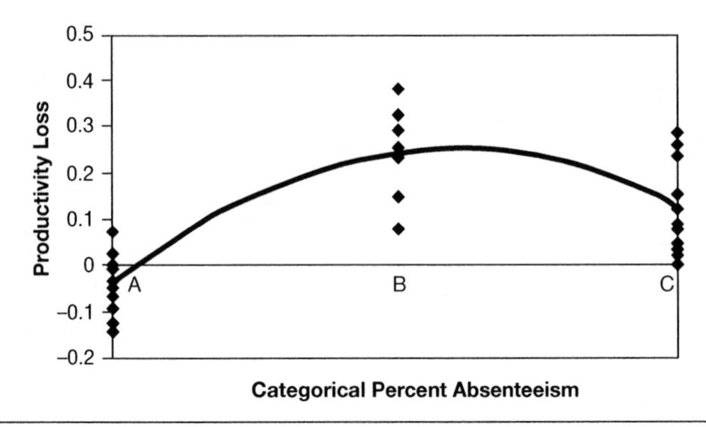

Figure 6.10 Productivity loss versus percent absenteeism.

Contractors reported turnover as a percentage of the number of replacement workers to the number of workers employed. The turnover data are more dispersed than the absenteeism data, chiefly because we placed a greater emphasis on absenteeism for this research. Our analysis on percentage turnover was also confined to three categories: 0–10%, 11–20%, and more than 20%.

Productivity loss was measured as the difference between budgeted and actual contract hours. The average productivity loss for the projects analyzed was 9.13% with a minimum of −14.44% and a maximum of 37.92%. We saw a wide distribution of values around a median of 7.27%. Negative values indicate projects completed in fewer than the budgeted contract hours.

We developed two statistical models for the quantitative macroanalysis. To describe quantitative relationships between productivity loss and the percentages of absenteeism and turnover, we used regression techniques. For each analytical test, we used a minimum confidence level of 90% to determine if a test was significant.

Absenteeism Model

We designated "productivity loss" as the response, or dependent variable, and percent absenteeism as the independent, or predictor variable. Then we developed the final model using ordinary least squares regression:

$$PL = -0.716 + 0.876(A) - 0.198(A)^2 \tag{6.1}$$

where:

PL = Productivity loss
A = Percent absenteeism

Table 6.9 Productivity Loss for Each Absenteeism Category

Level	Percent Absenteeism	Productivity Loss (%)
A	0–5%	-3.8
B	6–10%	24.4
C	More than10%	13.0

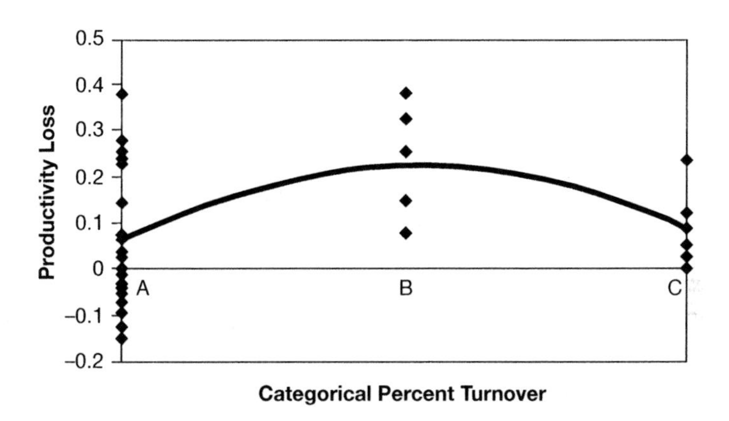

Figure 6.11 Productivity loss versus percent turnover.

Figure 6.10 graphically depicts Equation 6.1 together with all data used in this analysis. The x-axis contains the absenteeism data, and the y-axis contains the productivity loss in decimal form. In other words, a productivity loss of 0.1 means a 10% difference between budgeted and actual contract hours. As noted earlier, we analyzed project absenteeism data in three percentage categories: 0–5%, 6–10%, and more than 10%. Table 6.9 shows the productivity loss at each level of absenteeism.

Note that productivity losses are highest at Level B (6–10% absenteeism) and then start to decrease. We are not sure why this is so. One explanation might be that contractors budget resources more efficiently when they expect to see higher rates of absenteeism. They are thus able to keep productivity losses down even though absenteeism rates increase.

Traditional statistical analysis indicates a strong regression model. For example, the adjusted R^2 for the model is 60.2%. We also used hypothesis testing to determine the significance of the independent variable coefficients in Equation 6.1. This test indicated that the coefficients were statistically significantly different from zero and applicable in explaining a relationship with productivity loss. To determine the overall value of Equation 6.1, we analyzed the diagnostic plots from the regression equation. All plots were verified as satisfactory.

Table 6.10 Productivity Loss for Each Turnover Category

Level	Percent Turnover	Productivity Loss (%)
A	0–10%	6.0
B	11–20%	22.2
C	More than 20%	9.0

Turnover Model

Our model for "productivity loss" using percent turnover as the predictor is similar to the absenteeism model. However, the turnover model is much weaker. Therefore, although we have included the statistics for the model, we advise caution in using them. We developed the final model to explain productivity loss caused by turnover using ordinary least squares regression:

$$PL = -0.396 + 0.603(T) - 0.147(T)^2 \qquad (6.2)$$

where:

PL = Productivity loss
T = Percent turnover

Figure 6.11 graphically depicts Equation 6.2 together with all data used in this analysis. The x-axis includes the percent turnover data, and the y-axis contains the productivity loss in decimal form. As noted earlier, we analyzed project turnover data in three percentage categories: 0–10%, 11–20%, and more than 20%. Table 6.10 shows the productivity loss at each level of turnover.

Note that productivity losses for turnover, like those for absenteeism, increase until Level B (11–20% turnover) and then start to decrease. Once again, we are not sure why this is so. One explanation might be that contractors budget resources more efficiently when they expect to see higher rates of turnover. They are thus able to keep productivity losses down even when they experience higher rates of turnover.

The adjusted R^2 for this model is only 11.2%, which shows small variability explanation. We used hypothesis testing to determine the significance of the independent variables in Equation 6.2. The tests indicate that the coefficients were statistically significantly different from zero and moderately applicable in explaining a relationship with productivity loss. To determine the overall value of Equation 6.2, we analyzed the diagnostic plots from the regression equation.

Summary

When we analyzed the data for ongoing projects (microanalysis) we found that absenteeism had a significant effect on labor productivity. The results for turnover were statistically somewhat weaker. Because of this and because of the small sample size, we advise limiting use of this research to companies with characteristics similar to those in this study.

Our analysis of the effect of both absenteeism and turnover on productivity losses produced similar bell-shaped models for each. In both cases, we saw that productivity losses increased to a midpoint and then started to decrease. Thus, the higher levels of absenteeism and turnover actually resulted in lower productivity losses. We are not sure why this is so. However, we believe there may be at least one explanation. Based on their experience, contractors may have expected to have higher rates of absenteeism or turnover on certain projects. By planning for these losses, they were able to use resources more efficiently and therefore have lower productivity losses in spite of higher rates of absenteeism or turnover.

RECOMMENDATIONS

Those who completed our questionnaires as well as those we interviewed provided many suggestions for reducing absenteeism and turnover. We have compiled their suggestions into the following categories:

- Redefine overtime
- Use incentive programs
- Change work schedules
- Establish safe worksites
- Become a company of choice
- Record attendance publicly

Redefine Overtime

We learned that it is common practice for workers to miss a day of regular pay to work hours that pay time-and-a-half. Although they have taken a day off, they still work a 40-hour week, but get overtime pay for any hours beyond regular work hours. Such workers, in effect, have given themselves a raise.

Some companies have redefined overtime in their labor agreements as a way of addressing this problem. We found the following two examples in our research:

- Eight (8) hours work with thirty (30) minutes for lunch, between the hours of 6:00 a.m. and 6:00 p.m. shall constitute the workday. Five (5) such days, Monday through Friday, shall constitute the workweek.

Saturday may be used as a makeup day, and, if utilized, a minimum of eight (8) hours must be scheduled.

- All hours worked in excess of ten (10) hours per day or forty (40) hours per workweek shall be paid at one and one-half (1.5) times the straight-time hourly rate. All hours worked in excess of sixty (60) hours per workweek, or on Sunday, or on holidays (including Easter) shall be paid at two (2) times the straight-time hourly rate.

This type of language has been very successful in getting workers to work 40 hours per week. However, both management and labor had to compromise on the details. In one case, the union wanted double time for any hours worked beyond 60 hours. Workers then agreed that they would not get time-and-a-half until they had worked 40 hours. In the other case, management agreed to raise the pay rates of journeymen and foremen and add two paid holidays and a week's paid vacation to the benefits package.

Our research also found that companies should not offer overtime as an incentive to keep workers from moving to other jobs. This approach can backfire and will adversely affect budgets.

Use Incentive Programs

We received mixed responses about incentive programs. Most of the workers claimed these programs would help reduce their absenteeism and turnover. However, management believed that incentive programs do not work. They felt that workers already receive good wages.

Although we did not study the effectiveness of incentive programs in this study, we did receive many ideas from respondents. Following are some of the responses we received:

- On larger projects, all employees who work 40 hours per week for all periods throughout the month are eligible for a drawing to win a prize at the end of the month.
- Every week, if 40 hours are worked the previous week, a ticket is earned. Every week during a large project, small prizes are up for lottery by drawing a ticket. At the end of the project, the accumulated tickets are put into a hat, and one name is drawn for a new vehicle or other large prize.
- On large projects, workers would get a raffle ticket every day they show up on time and show a clear safety record. At the end of every month, have a raffle. Then at the end of the job, raffle a high-priced item. This would promote safety and reduce tardiness and absenteeism.

- Workers should receive a reward of tools if 40 hours are worked on a consistent basis.
- A bonus (shared with apprentices) goes to the lead craftsperson for completing a project in fewer hours than estimated while maintaining a quality of work acceptable to the company. This program promotes teamwork.

Change Work Schedules

Workers reported that one way to reduce absenteeism was to schedule 40 hours in four days instead of five. A schedule of 10 hours per day, Monday through Thursday, would leave workers time on Fridays to conduct business or rest. A missed day under this schedule would also result in a higher financial loss for workers. Another suggestion was to schedule four 9-hour days with a half-day on Friday.

Establish Safe Worksites

Our research showed that injuries play a key role in absenteeism. Workers also reported wanting to work on safe construction sites. Potentially dangerous tools, equipment, and materials exist on all construction sites. Therefore, an ongoing, proactive safety program can significantly reduce both absenteeism and turnover. Having a good safety record can result in a company becoming a company of choice. We learned that a company of choice is one that proactively inspires worker productivity and loyalty and thus has fewer problems with absenteeism and turnover.

Become a Company of Choice

Many of the companies we surveyed said they did not have much of a problem with turnover simply because they were a company of choice. Companies of choice enjoy a stable workforce because their programs and policies build worker loyalty. Following are some of their methods:

- Provide safe worksites
- Treat workers with respect
- Provide lunch
- Host company get-togethers
- Promote an “open-door” policy between workers and management
- Provide training or assistance with training
- Provide daycare or assistance with daycare
- Provide flexible schedules

Record Attendance Publicly

Our research also turned up a very simple but effective way to reduce absenteeism. Supervisors can post the attendance record where all employees can see it every day. Workers can track their own attendance and also see clearly who is not showing up for work on a regular basis. This technique demonstrates that management values attendance and takes absenteeism seriously. Workers who have already missed some time may be reluctant to see more absences marked next to their names.

CONCLUSIONS

Construction work is deadline-driven and labor-intensive. As previous research has shown, between 40% and 60% of a contractor's cost is in labor. Therefore, it makes sense for contractors to know more about the problems that can decrease labor productivity. Also, the problems of absenteeism and turnover are often overlooked. Most contractors have neither the time nor the resources to effectively study the extent to which these problems affect their productivity. As most contractors are aware, too much absenteeism and turnover can adversely affect labor productivity. Finding ways to effectively deal with these problems is an industry-wide challenge. Even an effective program will become less so when there is plenty of work to go around. Of course, the behavior of some employees will never change, no matter what the company does. Yet, as our research indicates, it is well worth it to proactively address these problems.

Our research has several limitations. First, our study focuses on electrical contractors only. To understand other construction trades, we recommend expanding the research base. However, in terms of the specific factors that cause absenteeism and turnover, this study was quite broad. Second, even the most well-intentioned contractors rarely have enough time to focus properly on data collection. We recommend that future research efforts make emphasis on developing partnerships with contractors to gain access to worksites to have researchers collect the data themselves. This approach would ensure that the data are collected properly and daily throughout the research. Finally, we focused solely on absenteeism and turnover. However, these problems are not the only causes of lower productivity. Other variables such as change orders, material delays, overtime, and shift work also affect productivity.

This book has free material available for download from the Web Added Value™ resource center at *www.jrosspub.com*

7

NEGOTIATING LOSS OF LABOR EFFICIENCY

Dr. H. Randolph Thomas, *The Pennsylvania State University*
Dr. Amr Oloufa, *The Pennsylvania State University*

INTRODUCTION

One of the economic consequences of construction is the loss of construction labor efficiency resulting from various events that occur during the course of construction. Occurrences such as scheduled overtime, schedule acceleration, numerous changes, and congestion can seriously erode contractor profits.

Although the construction contract may grant the contractor economic relief for labor inefficiencies, recovery of these monies is not guaranteed for several reasons. The economic consequences of losses of labor efficiency can be quite severe. It is not unusual for a contractor to experience labor productivity losses in the range of 25% to 40%. Unlike material and some equipment resources, using additional labor resources cannot be easily related to some cause or event for which there is entitlement, and there are no invoices to establish a link. A common argument by owners and general contractors for avoiding payment is that the cause of the labor overrun is incompetent management on the part of the specialty contractor. When combined with the magnitude of the losses, it is little wonder that the owner or general contractor is reluctant to acknowledge that it has an obligation to pay.

The purpose of this chapter is to provide guidance for the contractor and subcontractor on presenting loss of labor efficiency claims. Following these guidelines will enhance the likelihood of recovering the additional labor costs. The emphasis in this chapter is on educating the owner or general contractor as to the causes and consequences of labor inefficiencies. Another essential emphasis is on establishing cause-effect relationships so as to substantiate the labor overrun. The purpose of this chapter is not to explain how to calculate damages.

Schedule delay damages and losses of labor efficiency go hand in hand. However, it is important to observe that a schedule delay analysis and a loss of labor efficiency analysis are not the same. Different analysis methodologies are applied, and the cause-effect relationships are very different. With a loss of labor efficiency, it means that it takes longer to perform a certain task. There need not be a work stoppage or time delay that is necessary to perform a schedule analysis.

Although loss of labor efficiency may result in delayed completion, loss of efficiency is not included as an element of delay damages. When permitted by the contract, both delay damages and losses of labor efficiency can be recovered. It is not considered double recovery to receive both types of damages.

MAKING THE CASE FOR LABOR INEFFICIENCIES

There are numerous events that can cause a loss of labor efficiency. Research at Penn State has led to the development of the Factor Model, which is graphically shown in Figure 7.1. This model is explained in this section as well as other issues that can make educating the owner about the causes and magnitude of work hour losses much easier.

The Factor Model

There are two broad categories related to the work that affect labor productivity. These are the work to be done and the environment in which the work is done. Both are shown in Figure 7.1. The figure shows that the inputs in terms of labor hours are converted to outputs or quantities of work through the application of some work method. The work to be done and the work environment categories can be viewed as either contributing to or inhibiting this conversion process. The work to be done is defined in the contract documents, and there is little that the contractor can do about this category.

Direct Factors Affecting Labor Efficiency

Of importance to the discussion on labor efficiency is the work environment. There are numerous factors that can affect labor efficiency. Fortunately, some of

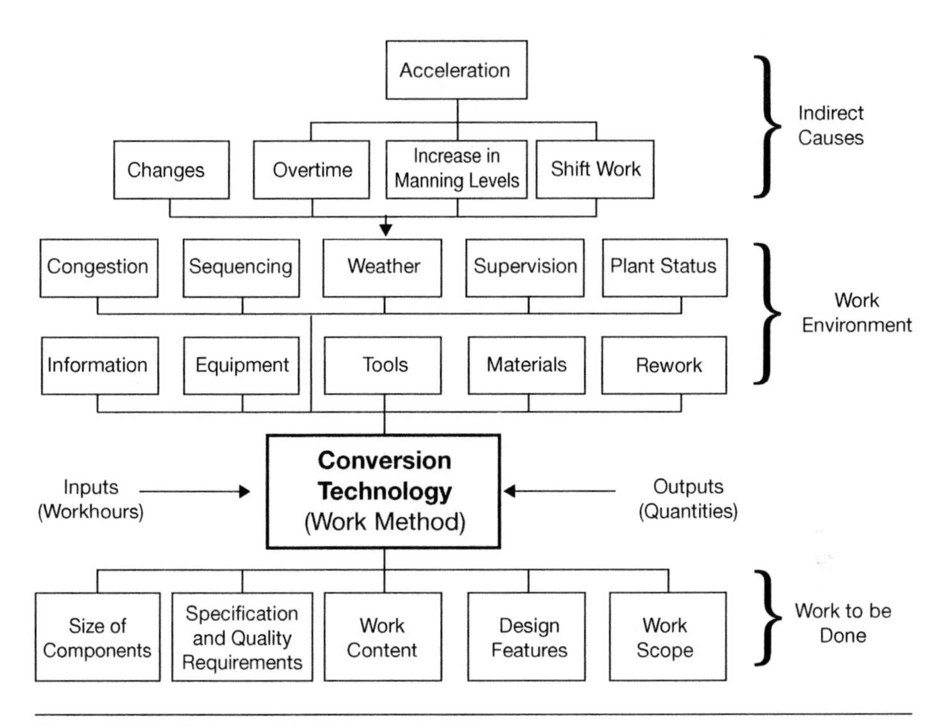

Figure 7.1 The factor model of labor productivity.

these factors are relatively minor and others generally occur infrequently. Research on more than 100 projects covering more than 4,000 workdays and 250,000 work hours of labor-intensive work has shown that there are only a few factors that consistently occur on large numbers of projects that negatively affect labor efficiency. These factors, which are also shown in Figure 7.1, are as follows:

- Congestion
- Out-of-sequence work
- Adverse weather
- Inadequate supervision
- Work performed while the facility is in operation
- Lack of information
- Lack of equipment
- Lack of tools
- Lack of materials
- Rework

These factors repeatedly occurred on the 100-project database mentioned above. By concentrating on these factors, arguments are more convincing because

Table 7.1 Impact of Disruptions on Daily Output

Factor (Disruptions)	Relative Daily Output
No Disruption	1.00
Congestion	0.30–0.35
Out-of-Sequence Work	0.20–0.30
Adverse Weather	0.35–0.50
Lack of Equipment	0.45–0.55
Lack of Tools	0.35–0.45
Lack of Materials	0.45–0.65
Rework	0.40–0.50

Table 7.2 Relative Output per Day

Day	Relative Output
Monday	1.00
Tuesday	1.00
Wednesday	0.30–0.35
Thursday	0.30–0.35
Friday	0.30–0.35
Average Weekly Output	**0.58–0.61**

there is reliable scientific research to support their effect on productivity. Research has documented that these factors have a significant effect on labor productivity (Thomas et al. 1990). Table 7.1 summarizes the relative daily output when some of these factors are present.

The results of Table 7.1 are applicable to the daily crew output. For example, on days when rework is performed, the average output is reduced to 0.40 to 0.50 of the normal output experienced when no disruption factor is present. Factors that occur most often relate to the lack of materials and weather. The most severe disruption factor is having to perform out-of-sequence work.

To illustrate the impact of the factors in Table 7.1 on labor productivity, suppose an electrical contractor is performing work in an area of limited space in a new commercial structure. The work took five days to complete. However, after the second day, the prime contractor changed the schedule and assigned two other specialty contractors to work in the same area. Thus, three days of the five were affected by congestion. The electrical contractor spent 225 work hours in this area during the five days. Pertinent data are shown on Table 7.2.

The output of the crew was reduced to approximately 60% of normal. The change in schedule caused an estimated loss of labor efficiency of (225)(1.0 – 0.6) = 90 work hours.

Indirect Factors Affecting Labor Efficiency

As shown in Figure 7.1, there are also indirect factors that affect labor efficiency. The common indirect factors are:

- Changes and change orders
- Scheduled overtime
- Increased manning levels or overmanning
- Shift work

These factors are considered indirect because the factors do not automatically lead to inefficiencies. For instance, there are certain times when scheduled overtime or changes work can be performed without a loss of productivity. How then do indirect factors lead to labor inefficiencies? To illustrate, consider scheduled overtime. Overtime creates an environment where the pace of the work is accelerated. Everything accelerates. Materials will be installed faster, and unfortunately, the material supply network may not be able to cope satisfactorily with the accelerated pace (Thomas et al. 1995). Without materials, the efficiency of the workforce is impaired. Thus, scheduled overtime is inefficient when it triggers the direct factors to occur, in this case, lack of materials. The same analogy can be applied to the other indirect factors shown in Figure 7.1.

EDUCATING THE OWNER

The first aspect of educating the owner is to keep the discussion simple. Convince the owner that there are relatively few factors that adversely affect labor performance. Concentrate on the fundamental or direct factors listed in Figure 7.1.

If surveyed, a list of 50 factors affecting productivity could be easily developed. Unfortunately, such a list leaves the owner with the feeling that the contractor is helpless in managing the work. This is not the impression that one wants to create. One approach is to use Figure 7.1 to illustrate that there are relatively few factors affecting productivity, and under ordinary circumstances, these factors are manageable. These are the direct factors shown. Other factors that were caused by or are the responsibility of the owner, called indirect factors, actually caused the inefficiency by triggering the direct factors to occur. The following examples illustrate how this may occur.

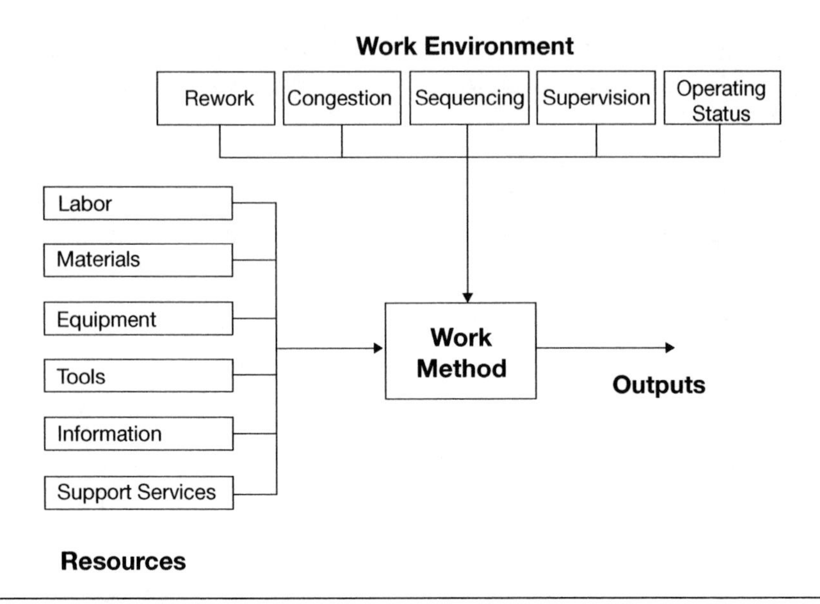

Figure 7.2 Flow diagram showing resources and environmental conditions.

Example 7.1: Consider the indirect factor of design changes, and suppose the owner has put certain work on hold pending a possible change. Some three months later, the owner informs the contractor that there will be no change and that the work should proceed as shown on the drawings. The environment at the work location is now entirely different than it would have been had the work proceeded three months sooner as planned. Now the contractor must work around other contractors performing other activities. The area is congested (stacking of trades) and different, less efficient methods need to be used. Some rework was also required because of damage from the other contractors. So, a relatively minor issue over a change that never materialized created congestion and the need for less efficient methods and rework that, in turn, caused a deterioration of labor productivity. The flow diagram in Figure 7.2 illustrates this simple concept. The contractor could argue schedule acceleration as a cause, but a more direct approach is to argue congestion and rework. The alternate method was caused by the congestion, so that argument may only confuse the simplicity of the situation.

Example 7.2: A contractor was contracted to install an underground fiber-optics innerduct across 60 miles of a northeastern state. Approximately 25 permits were required from numerous townships, the State Department of Transportation, and a major railroad. Unfortunately, the contract was awarded without any permits being secured. The contractor had areas of work made available to him in a piecemeal fashion. While it could be argued that the lack of permits was the main cause of the labor productivity overrun, the owner could counter that there was no work stoppage and there was always some work available for the contractor to do. A better position for the contractor is that out-of-sequence work was the main cause of lost efficiency. Out-of-sequence work resulted in the need to frequently change locations, find busy work to do when limited work was available, and return to work areas to complete tie-ins.

The focus of educating owners and general contractors should be on changes in the environment. These factors are the direct causes of lost productivity and have the greatest impact on how efficiently the work is performed. Because there are relatively few environmental factors, the discussion can be kept simple.

Common Indirect Factors

There can be a number of indirect factors; however, there are some that occur more frequently than others. These are explained below. Understanding why there is loss of efficiency is important because a clear understanding makes it easier to explain and convince the owner that his or her actions were the root cause of the labor overrun and it establishes the facts that are important to document, which creates the element of proof.

Design Deficiencies

The contract drawings show the intent prescribed by the designer. Specialty contractors use their knowledge and skill to efficiently construct the work within the bounds of this intent. When the design is deficient, the specialty contractor has difficulty in understanding what the designer wants. The specialty contractor must stop and wait for further instruction and interpret the drawings to his or her best ability, and sometimes the intent is innocently misunderstood. Each of these situations leads to a similar conclusion; that is, the specialty contractor must stop and wait for clarification, go elsewhere while the designer makes a determination, wait for new materials to be procured, perform rework, or a host of other scenarios. Each situation can lead to loss of labor efficiency.

Why are there design deficiencies? Two reasons seem to occur frequently. First, some designers have a tendency to overspecify what is required rather than letting the contractor use his or her ingenuity to get the job done efficiently. By overspecifying, discrepancies are often created, particularly with dimensions. The contract documents may then be uncoordinated leading to rework and out-of-sequence work. When the designer is behind schedule or has not been given adequate time to develop the contract documents, errors and omissions inevitably occur. The origin of these deficiencies can often be traced to indecision or procrastination by the owner.

Changes and Change Orders

Changes are often related to design deficiencies, but there can be other causes. These include changes in regulations, contractor errors, contractor requested changes, and owner preferences. Regardless, labor inefficiencies will result unless the change is identified early. The key determinant in labor inefficiencies resulting from changes is the timing of the change. If the need for a change is identified early, the impact is minimal or nonexistent. The absolute worst scenario is for the crew performing the work to identify that something is wrong and that a clarification or change is needed. As in the example described earlier, even a nonchange can lead to inefficiencies because the environment in which the work is done has changed.

Shop Drawings and Submittals

The review and approval of shop drawings and other submittals is a very important process that occurs early in the construction process. Contractors should expeditiously and carefully initiate this function. The contractor must take care in preparing a schedule of submittals to ensure that long lead items are done first and that all items are submitted in a timely manner. All too often, the approval process is delayed, and these delays ultimately lead to delays in the construction. The causes of delay are relatively few. Submittals may not be submitted on time or the submittals are incomplete. Both are the responsibility of the contractor. On other occasions, the delay may be caused by the owner or designer because the submittal is not returned in a timely manner. A rule of thumb is that the designer should be able to return the submittals to the contractor within two or three weeks. Obviously, owners and designers should avoid a cumbersome and lengthy review process. All parties need to be especially diligent where the design schedule is tight or has been delayed. Sometimes, designers will use the review process to buy time by disapproving the submittal as incomplete, not prioritized, or other reasons. Pay special attention to the submittal process. Do not give the owner reason to blame you for delays in the submittal process.

Late submittals need not involve your line of work. Frequently, other items such as structural steel may be the culprit. These delays are likely to lead to schedule acceleration impacts to the specialty contractor long after the steel erector has left the site. Thus, maintaining submittal logs is important. Other subcontractor's logs can also be relevant. This information may be readily available from minutes of the weekly progress meetings.

Late submittals lead to obvious delays in procuring materials and installed equipment. Less obvious is the implication when submittals are approved in a random sequence. This situation may mean that the work must be done out of sequence. The material procurement, storage, and distribution process may also be more difficult to manage.

Scope Increases

When there are increases (or decreases) in the scope of work, several situations can occur. If the scope changes are late, then the work will need to be done in a less favorable environment. If the facility is operational or occupied, then the work will be impacted further. It may be more difficult to obtain materials, and there may be premium charges added by the vendor for quick deliveries. Another problem that can occur is that the scope change can involve smaller segments of work than would have been ordinarily required. Here the contractor must distribute additional setup time and costs (fixed cost) over a much smaller quantity of work. The work environment may also be more difficult. Obviously, it is important to identify scope changes early in the construction process to avoid significant disruptions to the work.

Schedule Acceleration

Schedule acceleration has been studied extensively in Chapter 4. It is fair to say that schedule acceleration is very inefficient and expensive. Labor inefficiencies over the total work have been calculated as high as 50%. Inefficiencies in the range of 25% to 40% can routinely be expected. Schedule acceleration leads to labor inefficiencies in a number of ways. One of the greatest impacts is in material management. The problem with schedule acceleration is that resources are consumed at an accelerated pace. Materials must be installed faster, equipment will be needed more frequently, and design issues and RFIs will occur more regularly. There may be limited space in which to store materials, and the distribution of materials into the facility may be more difficult. The management of waste materials and trash is another potential issue.

In acute acceleration situations, the ability to plan ahead is impaired because there are so many schedule changes that it is hard to say where a crew will be working the next day or week. The orderly sequence of the work is disrupted, leading to out-of-sequence work. The planning horizon becomes very short. Where scheduled overtime is used extensively, fatigue may become a problem. Where overmanning occurs, there will be dilution of supervision and stacking of trades.

Cause-Effect Diagram

Obviously, educating the owner is not a quick and easy task. It takes some thought because the owner does not want to pay for the labor inefficiency cost, which can be substantial. Your best chance for success is to keep it simple. However, recovery is predicated on how well you show the cause-effect relationships. There must be a showing that what the owner or general contractor did, caused you to take

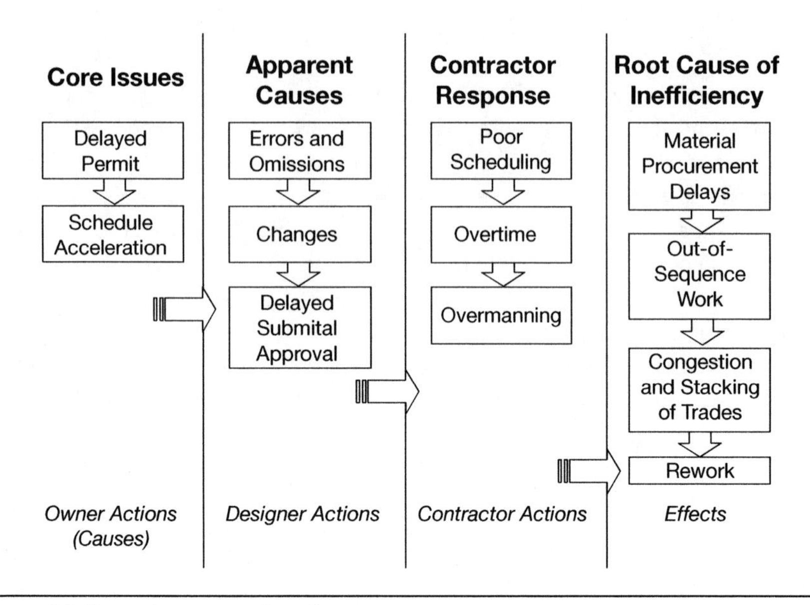

Figure 7.3 Illustrative cause-effect diagram.

certain actions, and those actions resulted in labor inefficiencies. You had no other alternatives and the contract entitles you to recover the added cost.
A cause-effect diagram like the one illustrated is Figure 7.3 may be helpful. In this case, the owner is late in obtaining a needed permit, which delays the start of the work. The owner accelerates the schedule by refusing to grant a time extension.

The indirect factors like changes or design deficiencies are the events that happen on the job. Make the connection between the indirect causes or events and the factors that actually cause labor inefficiencies, like lack of materials, out-of-sequence work, or rework.

Returning to Example 7.2,a cause-effect diagram is illustrated in Figure 7.4. Notice the simplicity of the presentation and how the causes of inefficiency are traced back to the failure of the owner to procure the necessary permits. In this situation, the owner reacted to the difficulties on the project in such a manner as to make the inefficiencies worse. These reactions are also shown in Figure 7.4. Figure 7.5 shows the work hour inefficiencies that occurred. Notice that in the first half of the project, the inefficiencies occur in groupings associated with the issuing of permits. This graphic clearly shows the link between inefficiencies and the lack of permits. The inefficiencies occur because of insufficient work to perform and moving to new locations. In the latter half of the project, the inefficiency patterns are different. These inefficiencies occur because of cleanup and tie-in issues and because removing work from the contract made it more difficult to maintain any orderly sequence to the work even on a daily basis.

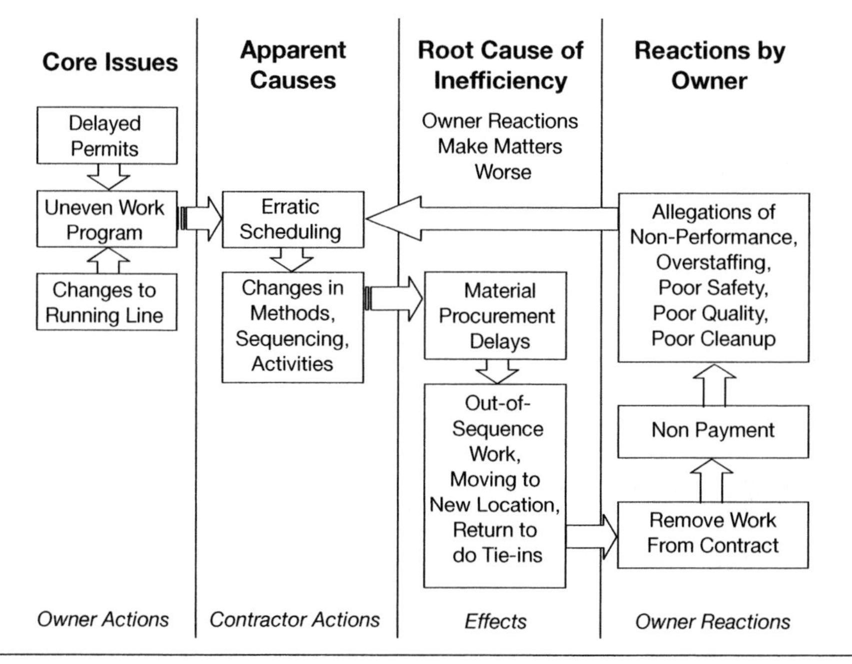

Figure 7.4 Cause-effect diagram for Example 7.2.

WHAT TO DOCUMENT AND WHEN

The chronology about losses of efficiency and how monies were lost is not an easy story to reconstruct, especially after the fact. Therefore, it is imperative that the documentation of events and conditions is done as the project is in progress, not later.

Contractors should organize a project file containing correspondence and other documentation. Documents are needed to show adherence to the contract requirements and to establish the factual basis of the claim.

Organizing a Project File

The documents and records needed to support a claim for a loss of labor efficiency form part of the overall documentation that collectively is called the project file. McDonald (1989) cites the checklist of 10 essentials to include in the project file:

1. *Estimating and bidding files.* These files should include the original estimates and related backup sheets. There may be other information that is helpful to substantiate the validity of the estimate.

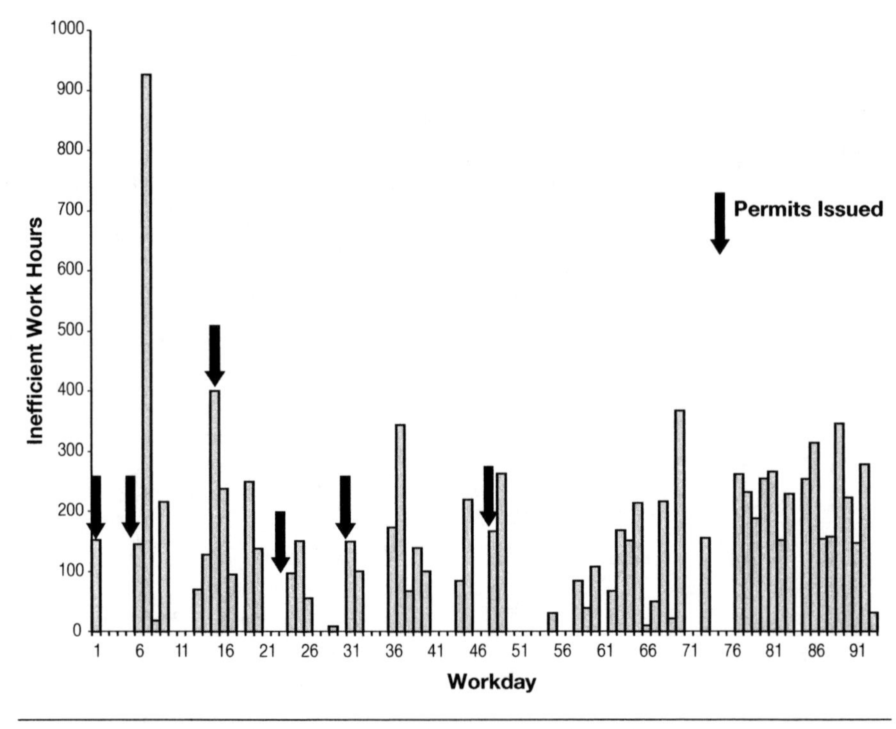

Figure 7.5 Inefficient work hours associated with the lack of permits.

2. *Complete file of contract documents.* This file should contain the entire contract including addenda, change orders, and any correspondence related to contract negotiations.
3. *Cost records.* The week-by-week cost records should be maintained. Additionally, all documents related to requisitions, deliveries, and payments need to be filed.
4. *Schedules.* This portion of the file should include preliminary, original, updated, and revised schedules.
5. *Correspondence and similar documentation.* All correspondence, internal memoranda, notes of phone conversations, minutes of meetings, and other documents are crucial in establishing the proof of key events and the timing thereof. Many claims fail because of the inability to establish when certain events happened or that they happened at all.
6. *As-built data.* Files should be maintained of all daily reports, inspection reports, daily timesheets, job diaries, inspector's and engineer's reports, and so on. These records may be the only documentation of the conditions under which the work is done. The job diaries are par-

ticularly important, and each foreman should maintain his or her own personal diary. It should be completed daily and include the nature and location of the work, quantities installed, visits and inspections, and relevant factors or events affecting the crew's performance.

7. *Standard form correspondence.* Files should be kept of phone conversations, RFIs, field clarifications, transmittals, and submittals. Maintaining transmittal, submittal, and changes logs should be a routing part of every contractor's operation.
8. *Subcontractor files.* While records of other contractor activities may seem irrelevant, there are times when it is important to be able to document their progress. Manning levels and the location of their workforce are important elements of information to substantiate out-of-sequence work or that the slow progress or interference of others affected your work.
9. *Photographic documentation.* Photographic records can be an important element of proving entitlement to a claim or to show that the conditions under which you worked were much different than you could have anticipated when you submitted your bid. However, the value of the photos is dependent on the care with which details are recorded. Photos should be used to substantiate disorderly material storage, random adherence to sequence by others, poor housekeeping, and other factors affecting the work. Panoramic "vacation-type" photographs are of limited value because of the absence of detail.
10. *Job completion documents.* This part of the job file should contain punchlists, certificates of substantial completion, certificates of occupancy or certificates of final acceptance. These documents and any protest you may file are important to show how others adhered to the contract documents in closing out the project. It should be noted that problems often occur with project closeout. From a contractual viewpoint, the warranty period begins when substantial completion is achieved, so establishing this date is important.

Documentation

There are two forms of documentation with each serving a specific purpose. The first involves correspondence and communication which are used to preserve contractual rights. This correspondence involves notice requirements, reserving rights to impact costs, responses to orally directed changes, and other forms of communication. Responses to orally directed changes and extra work are particularly important. Always make your position clear using language that is concise and straight to the point. Avoid the use of "weasel words." Specific language such as the following should be routinely incorporated:

> *It is my understanding that on [date], you directed me to perform certain work that is not called for in the contract documents. [Describe the work.] Since this is beyond the scope of our work and is extra, I am expecting that a written change to the contract is forthcoming. If my understanding is incorrect, please advise me by [date]. Otherwise, I will proceed with this work under the assumption stated above.*

Other elements of the communication involve making positions clear through correspondence and minutes of meetings.

The second form of documentation establishes the factual basis of a claim. This task involves maintaining logs of changes, RFIs, submittals, and other relevant items. Photographs and videos can be very effective in substantiating a variety of conditions and circumstances.

Correspondence and Communication

The contract defines various obligations of the subcontractor. Therefore, it is imperative that you read the contract before a problem arises. If a dispute arises later, there will be no sympathy from a general contractor, owner, arbitrator, or judge if you have not read, understood, and complied with the contract.

Notification of Possible Claim

All contract requirements must be followed. First and foremost, contractors must comply with the notice requirements of the contract. The time limit for compliance varies, so a careful reading of the contract is necessary. The form of compliance should be a letter to the contract administrator or to the person designated in the contract. The letter should not be accusatory or hostile. Rather, it should state the facts as you know them and indicate that it may become necessary for you to submit a claim at a later date. Always cite the relevant contract language that you feel entitles you to an equitable adjustment. Whenever there is doubt, write a notice letter. This is the only way to preserve your rights. Sadly, there are many instances where failure to file notice has precluded an otherwise valid claim.

Impact Costs

One problem that arises relative to asserting a loss of labor efficiency is that there may be impact costs. Impact costs are often associated with changes, but they can occur in many other ways. These costs are very difficult to estimate. For instance, work may have been bid in the contract at 1.0 work hour per unit. A change may be priced and negotiated by the owner or prime contractor at 1.2 work hours per unit. However, when the work is actually performed, it may take 1.7 work hours per unit. This additional 0.5 work hours per unit are impact costs and are often the result of rapidly deteriorating conditions at the site resulting in congestion, out-of-sequence work, and so on. These conditions may be difficult to foresee at the time the change proposal is made. On change proposals, reserve your rights to impact costs. Unless this is done, at a later date, you may not be able to assert entitlement to impact costs. How then do you reserve your right to impact costs when the change proposal says that 1.2 work hours per unit is the limit of entitlement? It is important that you not sign the change proposal unless you have added a statement that you reserve the right to file a claim at a later date for impact costs. Keep the owner and general contractor abreast of impact costs as you observe them. Understand that these costs may not become apparent until later in the project.

Meeting Minutes

The minutes of weekly progress meetings are an important project document. It is your responsibility to make sure it contains a fair and accurate record of difficulties affecting your work. Therefore, you must first go to these meetings prepared to discuss problems and with the willingness to support these with notes, dates, correspondence, logs, and so on. Minutes may also serve as a form of notice. Insist that minutes be published in a timely manner. Once minutes are published, read them carefully and offer corrections in writing of any misstatements of a factual nature. Unless this is done, the minutes may be accepted at a later time by an arbitration board as being accurate. Keep copies of all such correspondence in the project file related to the minutes.

Documenting the Facts

Contractors need to be judicious in how a claim is substantiated by the facts. It is imperative that one be able to establish who did what to whom and when it happened. Being able to do so can require many forms of documentation that are too numerous to list. However, a few suggestions may be helpful. Maintain up-to-date logs of submittals, change proposals, diaries, and other relevant data. These logs are an important source of information to establish a timeline of important events. Logs are an important source of factual information. These are simple records of when project documents are transmitted and received. The most common logs are of submittals and change proposals. Logs should include a unique numbering system for each record and should contain the dates of transmittal and receipt. Also, there should be a comment column to note conditions of acceptance or reasons for rejection. There should also be a cross-reference to any correspondence related to the log entry. The types of logs can be numerous and specific to a particular job. Other types of logs include telephone conversations, contractor directives, and a record of visitors.

PRICING AND TRACKING CHANGES

Changes are an integral part of any construction work. How these are priced and tracked can make the difference between being paid or not. Much has been written about pricing and tracking changes. The following briefly reviews some of the more important points as they relate to loss of labor efficiency.

Pricing Changes

Changes can vary widely in scope and timing. Changes can be additive, deductive, or have no impact on the construction work. Some may even be for the convenience of the contractor. It is important for the contractor to take the first change as seriously as the last change. Usually changes evolve as a number of small changes, and it is the cumulative effect that eventually leads to significant losses of labor efficiency. If the change is a large one, the impact can be anticipated. However, with small ones, the impacts will be gradual and may not be noticeable until near the end of the project.

Pricing the Conditions

There are three levels of estimating the labor hours needed for a change. These levels relate to differences in the environment in which the change is to be performed. For convenience, these levels are described as green, yellow, and red. These levels are analogous to normal, abnormal, and severe conditions.

Green (Conditions Normal)

When conditions are normal, conventional estimating techniques, manuals, and databases can be used. These resources can provide a reasonable estimate of the hours needed to perform the work.

Yellow (Conditions Abnormal)

Many factors can cause conditions to become abnormal or difficult. Among these are severe weather, congested conditions, rework, having to work in an operating environment, and others. Whenever one or more factors cause the conditions to change, conventional estimating methods will not be satisfactory because conventional methods are based on normal conditions. The most common way contractors handle abnormal conditions is to identify a number of impact factors. An example of impact factors for one contractor performing electrical and mechanical work is included in Table 7.3. The estimated work hours determined from conventional estimating methods are multiplied times the appropriate impact factors. Generally, only one or two factors are applied to a single change. If multiple factors are present, then this method may lead to faulty estimates. Using these factors, contractors can usually arrive at a reasonable estimate. The problem, however, is that the owner does not want to pay for a change with what appears (to the owner) to be an "inflated" labor rate.

In Table 7.3, the conditions are rated and the total percentages are summed. The "normal" estimate is multiplied times the percentage to reach the final estimate. For example, suppose a change will be done where there will be congestion and stacking of trades. The percentage increase from Table 7.3 is 5%. Therefore, if under normal conditions, the change is estimated to require 375 work hours, then for the conditions described, the change should be estimated to take 375 (1.05) = 394. If materials can be located no closer than 250 ft, then the work hour estimate will increase to 375 (1.08) = 405.

Red (Conditions Severe)

Condition red is the most difficult situation for pricing changes because there are no methods for producing reliable estimates other than experience. In the red condition, the sequencing of the work and the project schedule is no longer followed. The schedule is usually accelerated and congestion, overtime, and overmanning are common.

The impacts on labor productivity are acute when conditions are severe, and it is unlikely that an owner or prime contractor would agree to such a change, even if a reliable estimate could be produced. Therefore, it is essential that in every change proposal, contractors reserve their rights to claim impact costs at a later date when actual costs are known. Two other aspects are important in the red or severe condition: (1) Keep the owner or general contractor fully informed of the factors affecting your work. Concentrate on the principal factors of sequencing, scheduling, congestion, overmanning, and so on. Keep a positive attitude in conveying these concerns. Try to be cooperative in getting the job finished without compromising your rights. Legal counsel may also be helpful. (2) Document additional cost as much as possible. Sadly, many claims fail because there is little to substantiate that the additional labor costs occurred because of factors beyond the contractor's control. Photographs are invaluable for collaborating losses of labor efficiency. The photographs should be made for the specific purpose of documenting the conditions in which the work is done. In this regard, panoramic photographs are of limited value. Some forethought should be given to the content of photographs.

Table 7.3 Impact Factors for Changed Work

	Scenarios and Adjustments							
Impact Factor	**Normal**	**Max (%)**	**Degree 2**	**Max (%)**	**Degree 3**	**Max (%)**	**Degree 4**	**Max (%)**
High Temperatures	None	0	None	0	Above 90°F	2	Above 110°F	3
Design	Complete	–2	Normal	0	Field Direct	10		
Material Handling	100'–200'	1	200'–300'	3	300'–500'	5	500'–750'	8
Congestion	Clear and Open	0	< 20 ft^2/person	5				
Height	0'–12'	0	12'–20'	1	20'–40'	3	40'–60'	5
Operating Environment	None	0	In Operation	4				

Tracking Changes

Tracking changes begins before the first change is issued. It is important that all changes are tracked, even those you are not sure are changes and others for which the owner or prime contractor denies a change order.

Logs and files of correspondence, RFIs, and submittals should also be maintained. These should be kept in a file such that all correspondence and paperwork associated with a particular change should be kept in a single folder. The filing system should anticipate that certain changes will not be made or will be denied. Others may be combined into a single larger one so the filing system must be flexible.

Changes File

A changes file should be maintained for every known or contemplated change. This file consists of a folder for each potential change. The folder should be labeled according to a unique numbering system. The folder should contain all worksheets, diaries, notes, quotations, invoices, observations, estimates, photographs, correspondence, schedules, sketches, drawings, and anything else that relates to the change.

Changes Log

A log should be maintained that tracks the dates of transmission of change proposals, responses, change orders, and other related correspondence. At a later time, these dates may be important in establishing the project timeline. Similar logs should be maintained for RFIs and submittals.

OWNER AND DESIGNER EDUCATION

The owner, designer, and general contractor must be informed early in the process that labor productivity is being affected. It is important that these communications not be threatening or accusatory, but rather, state the facts as you know them.

Preconstruction Actions

Effective preconditioning of the owner, designer, and general contractor requires that the specialty contractor present a positive and constructive attitude. Professionalism is important to maintain open lines of communication, and open communication begins at the outset. Before construction work begins, contractors should do the following:

- *Establish contractual lines of communication.* For each of the primary parties, identify the person who has the authority to make things happen. Always follow the command chain. If you cannot get satisfactory resolution from the next person up the chain, request a meeting where all relevant persons are in attendance.
- *Determine the time schedule for submittals.* Many jobs start on the wrong foot because the submittals are not submitted or reviewed in a timely manner. An overall time schedule should be established for various submittals, and all relevant parties, including the owner, should agree upon this schedule. Make sure submittals are submitted promptly and as thoroughly as practical. Do not let yourself be blamed for delays in the submittal process.
- *Involve the owner in the submittal process.* It is in everybody's interest to have submittals reviewed thoroughly and returned promptly. The owner should be keenly aware of and monitor this process, and it should be discussed at all weekly progress meetings until the reviews are complete.
- *Communicate a summary of logs to the owner.* The owner should routinely be advised of the status of submittals, changes, and RFIs. Periodic meetings should be held as needed with all affected parties just to discuss these issues.

Construction Execution

During the construction phase, contractors should do the following:

- *Keep lines of communication open.* Communication with the general contractor and owner is very important. Make sure that the owner becomes a participant in all important meetings. Make sure the minutes of meetings are accurate, and keep correspondence professional and factually correct at all times.
- *Use linear scheduling.* CPM schedules are very detailed and can be difficult to follow. A problem that occurs frequently, especially on commercial and institutional projects is that the sequencing of the work becomes inefficient. That is, instead of the work progressing from one floor to the next, work progresses on all floors simultaneously and the completion of the project is not orderly. This situation is not easy to show visually using CPM, but it can be readily seen using linear scheduling methods. Use this graphical scheduling method to show owners how the work is out of sequence and explain how this impairs labor efficiency. Figure 7.6 is an example of a planned sequence for a

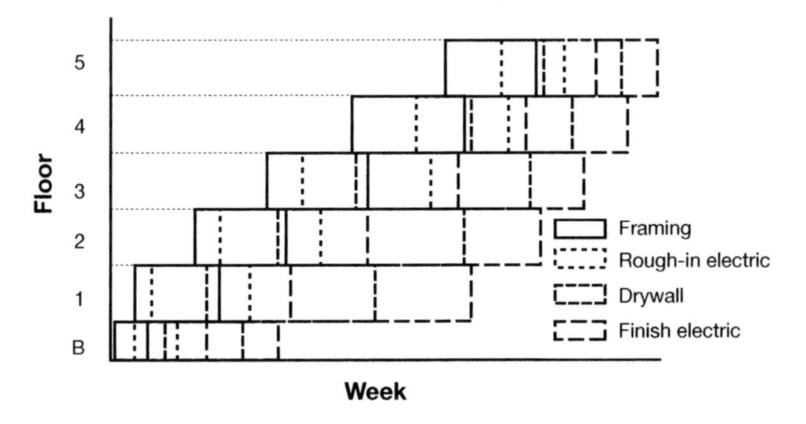

Figure 7.6 Sample linear schedule for a five-story building showing planned sequence.

five-story building. Four activities are shown. As can be seen, work is planned to progress from one floor to the next in an orderly progression, and completion of the work occurs in a sequential manner.

- *Consider using outside consultants.* When jobs become chaotic, short-range planning suffers. The conditions gradually deteriorate, and the symptoms of planning deficiencies begin to appear. While these symptoms can easily escape the contractor's notice, they are readily obvious to the trained eye. Outside consultants can identify conditions related to the work that, if removed, can make the work more efficient and minimize the potential economic losses. If a report is prepared, it can be used to convince the owner that labor inefficiencies actually occurred. Such reports are much more effective if done during the course of the work rather than later when a claim has been filed. If receptive, engage the owner as a participant.

Post-Construction

Once the project is finished, there is little that can be done to educate the owner aside from an effective claim presentation, which is discussed in the next section. However, throughout the post-construction period, contractors should be responsive to inquiries by the owner and should always be courteous and professional.

CLAIM PRESENTATION PRINCIPLES

An effective negotiation session and claim presentation should follow certain principles. The principles cited below are general and each claim will be unique.

The next section is a case study that illustrates how this uniqueness can be expressed.

General Principles

A positive attitude is important to convey to the owner that it is economically advantageous to the owner to settle the claim because you, the specialty contractor, are going to prevail if the dispute goes further. Obviously, this is no time to try to bluff your way to settlement. How you present the facts will convey the strength of your position. Therefore, you must go to the core issues that are the strength of your claim. These issues should be relatively few and logical. Also, be sure to remove emotion from the negotiations. In preparing for the claim presentation and negotiating session, it may be helpful to identify the person on the other side that seems to understand the facts and is understanding of your position. If you can convince that person, he or she may be able to convince the others.

Lastly, where possible make use of previous court cases that support your entitlement to damages. The circumstances should be similar to your situation. The cases need not be from your own trade, since the contract language and surrounding circumstances are what is important, not what is being built.

Presentation Principles

In making the actual claim presentation and in subsequent negotiations, it is essential to establish cause-effect relationships. There must be a showing of what the owner or general contractor did, how that changed your work, how you responded, and how and why the situation resulted in the less efficient use of labor resources. If you cannot establish cause-effect relationships, you have little or no chance to recover monetary damages. Instead, it will appear as if you are merely complaining.

The presentation must be factually correct and logical. It should be free from superfluous facts and issues. The entire purpose of the presentation is to tell a story. The presentation needs to be a "Dick and Jane" story. It should focus only on relevant facts and circumstances and should concentrate on the core issues. Above all, it must be presented logically and it must be straightforward and simple.

The person with the authority to agree to a settlement probably was not intricately involved with the project. This person will not be able to easily follow a detailed presentation, especially if the facts are not central to the claim. The claim presentation should follow these principles:

- The presentation should be supported with documents such as letters, logs, minutes of meetings, and photographs. Highlight the relevant parts of documents so others can readily find them. Be conscious of information overload. Too much information will only make the timeline difficult to follow.

- Supplement the presentation with simple charts, graphs, and summaries. These should be your own, since project documents are often too detailed to follow. Graphs and charts should summarize important facts and leave out unnecessary details.
- Give careful consideration in selecting the person to make the presentation. Quite often, a person closely involved with the project is not a good choice because their emotions may come across too strongly. Further, the tendency to elaborate and provide irrelevant details may be too great for the presentation to be effective.
- When conducting the meeting, follow a prepared agenda. This is your meeting and you should be in charge. Avoid discussing items that will be discussed later in the meeting.

Negotiating Principles

Negotiations need not be intimidating provided several principles are followed to avoid a confrontational environment. Anderson (1992) outlines four points dealing with elements of negotiation:

1. *Separate people from the problem.* It is necessary to separate the substantive elements of the claim from personalities and the relationship between the parties. Personal attacks will make it difficult to reach any kind of equitable settlement.
2. *Focus on interests, not positions.* It is easy for egos to replace the interests of the parties. The result is that the positions of the parties can mask what the participants really want or need. It is often more effective to focus on the underlying need to be made whole, which is the real cause for the company taking a certain position.
3. *Maintain flexibility.* It is difficult to arrive at satisfactory settlements under pressure and in the presence of an adversary. Keep an open mind as to settlement options, and know your flexibility prior to the negotiating session.
4. *Insist on objective criteria.* Discuss the conditions of the negotiation in terms of the language of the contract. By doing so, neither party needs to give in to the other, and both parties defer to the fair and equitable solution established in the contract.

Preparation for negotiation is also important. Following are several rules to follow that can increase the likelihood of a successful negotiation:

- *Do your homework.* This involves hard work and discipline to understand all the facts and focus on the core issues.
- *Go to the top.* You must hold discussions with persons who are authorized to make an equitable adjustment. Do not rely on someone else's ability to communicate the story on your behalf. In addition, dealing with others who cannot make contract adjustments lessens accountability and may hinder the proceedings.
- *Avoid quick concessions.* Make concessions only after considering all the issues and risks.
- *Accentuate the positive.* By framing negative points in a positive way, you will be more likely to elicit a positive response. Couching controversial issues between positive points increases your chances of getting others to listen and agree with you.
- *Maintain your composure.* Keep the discussion on the facts without getting personal. Make sure all statements made contribute positively to a desirable outcome.

CASE STUDY

In this section, a case study of an actual arbitration is presented where an electrical contractor brought the arbitration action against the general contractor. The purpose of this case study is to illustrate how you may create documents to tell the story of lost labor efficiency and how to establish cause-effect relationships. In the case study, labor inefficiencies were calculated following the methodology explained in Chapter 4. Therefore, how the calculations were made is not explained herein.

Project Description

This project involved the renovation and construction of a federal courthouse in the northeastern United States. The existing facility was a five-story building constructed in a U-shaped configuration. In the open area called the infill area, the height of the existing building was two stories. Storage areas for construction materials were very limited.

The construction work involved the demolition of all floors, leaving intact the structural frame, which was reinforced concrete. Some asbestos was present but played no role in the ensuing electrical claim. The infill area received a new five-story structural steel frame. This work consisted of erecting approximately 200 to 250 pieces of steel, an activity that should have taken two to three weeks. When completed, there was no longer a vacant infill area. New concrete floors were constructed, and all remaining work was consistent with the construction of a new facility.

The electrical contractor's rough-in work was integrated with that of the framing and drywall contractor. The millwork contractor followed the drywall work. The millwork was most detailed on the top three floors where the larger courtrooms were located. The paneling was rather expensive, and some of the finish work, including finish electrical, could not proceed until the millwork was finished. The electrical fixtures in the courtrooms were also very expensive. These could not be installed until all other work was completed. Other elements of the work, like the sound system, could not be tested until all work was completed. Therefore, the electrical contractor's work depended on other contractors performing their work in an orderly and timely manner.

Basis of Claim (Core Issues)

The work of the electrical contractor was impacted because various specialty contractors were late in performing their work, and when performed, it was done out of sequence and not prosecuted aggressively. Work was not completed on a particular floor before proceeding to the next. Thus, the electrical contractor was forced to complete all its work in the last few months of the project. Thus, the core issues in this dispute are as follows:

- The general contractor did not manage the subcontracts by allowing the steel erection, framing and drywall, and millwork contractors to work at a leisurely pace.
- The general contractor did not manage the project schedule by allowing the framing/drywall and millwork contractors to work in random locations rather than follow the floor-by-floor sequence established by the CPM schedule.

Compartmentalize the Analysis

A careful examination of the progress of the work indicated that the work evolved through four phases. Dividing a project into phases is an effective way to compartmentalize the discussion and, thus, simplify the telling of the story. This makes it easier for owners and arbitrators to understand. The various phases should be logically selected based on the conditions in which the work was performed and how the work was made available to the electrical contractor. In this case study, four phases were identified. These are summarized in Table 7.4 and shown graphically in Figure 7.7. Figure 7.7 shows the four phases in a diagram developed following the methodology explained in Chapter 4 where the actual and normal weekly

Table 7.4 Definition of Case Study Project Phases

Phase	Dates	Description
1	Sept. 4, Year 1–Dec. 31, Year 1	Steel Erection Delay: The infill steel erection was delayed by five months.
2	Jan. 1, Year 2–Apr. 29, Year 2	Framing-Rough-in-Drywall Sequencing: Infill steel was completed opening up significant areas of work. Rough-in electrical was sandwiched between framing and drywall.
3	Apr. 30, Year 2–Sept. 9, Year 2	Millwork Delay: The millwork contractor worked at a slow pace and out of sequence. He worked on all floors simultaneously and did not finish any single floor until the latter part of August, Year 2. The electrical contractor could do nothing but wait.
4	Sept. 10, Year 2–Nov. 4, Year 2	Rush to Completion: All other contractors completed their work and left the area to the electrical contractor. The fixtures, much of the finish electrical, and testing of various systems were completed in about five to six weeks.

labor percentages are shown. These data are used to calculate the number of inefficient work hours per phase.

By partitioning the project into these four phases, specific events can be limited to a particular phase, thus keeping the discussion focused on the core issues. The factors affecting labor efficiency were different in each phase. Table 7.5 summarizes the factors that affected labor efficiency during each phase. In this case study, the lengthy details are reduced to a few causes that can be readily understood.

Presentation and Arbitration Exhibits

The presentation during the arbitration proceedings was done by the phases listed in Table 7.4. The actual exhibits consisted on a collection of slides that explained the major issues. These exhibits were organized by phase and cover key events, nature of the work, causes of labor inefficiency, and the numerical estimate of inefficiencies. In this chapter, we have summarized those exhibits into Tables 7.6 to 7.12 and Figures 7.7 to 7.9.

In these exhibits the facts are condensed to only the essential matters that relate to the core issue, which is that the general contractor did not effectively manage the subcontracts or the project schedule. Contractually, this is the basis for the claim. Compartmentalizing and condensing facts is consistent with telling a "Dick and Jane" story. Notice also that the graphics are simple. Simplicity means

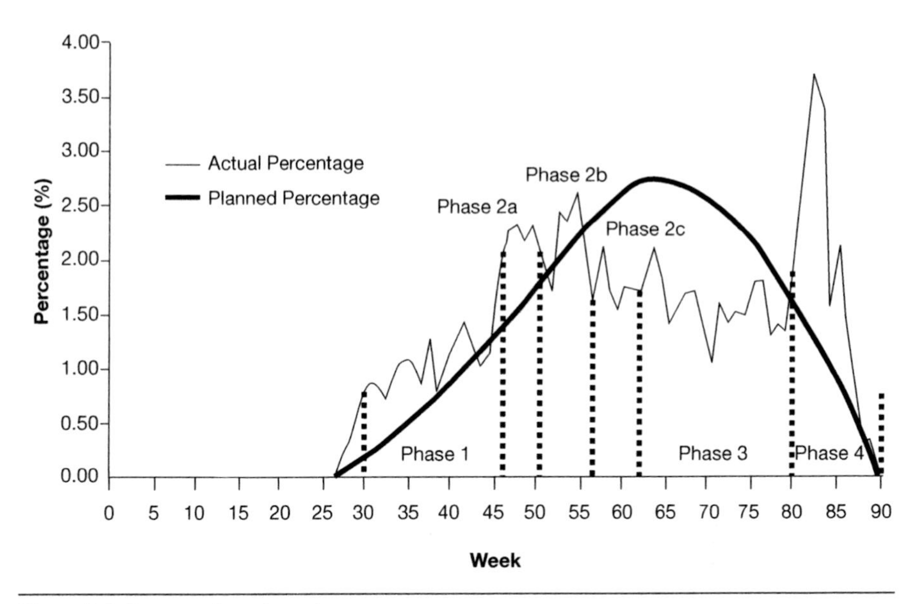

Figure 7.7 Case study project phases.

Table 7.5 Factors Affecting Labor Efficiency by Phase

Phase	Factors Affecting Labor Efficiency
1	Piecemeal and out-of-sequence work
2	Interference with framing and drywall contractors requiring inefficient methods
3	Little work available to perform; piecemeal work
4	Intense acceleration, overmanning, overcrowding, and out-of-sequence work

that there is no need for elaborate verbal explanations that would make comprehension more difficult.

In this case there are two core issues as shown in Figure 7.8. This figure also illustrates why the contractor is entitled to additional expenses associated with inefficient management by the general contractor.

Phase 1

The discussion of Phase 1 is covered in Table 7.6. There are only five key events, but these are sufficient to explain the effect that late steel erection had on the framing and rough-in electrical. The nature of work performed is explained considering that the infill area was off limits. Finally, the causes of labor inefficiencies

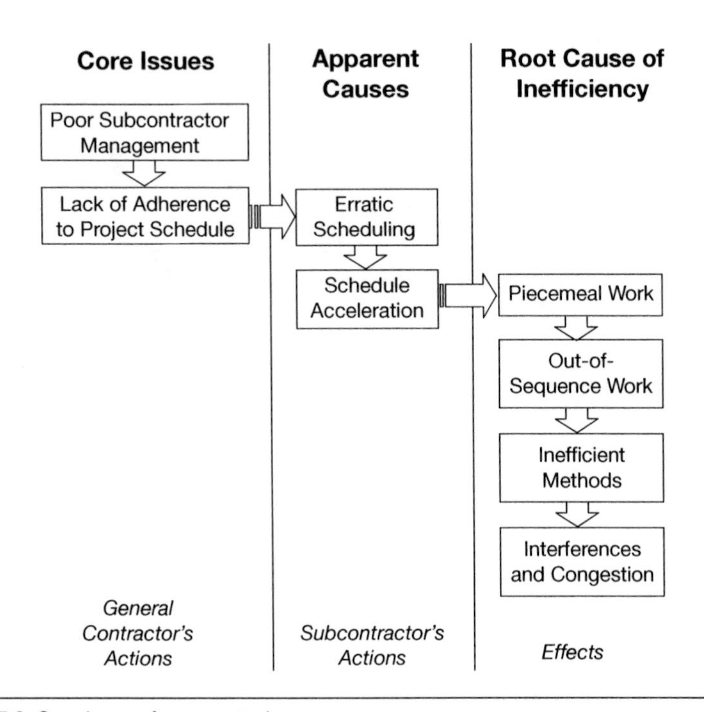

Figure 7.8 Core issues for case study.

are mainly piecemeal and out-of-sequence work. Notice how Table 7.6 reduces the facts to an essential few. It details the causes that are beyond the contractor's control, summarizes the contractor's response, and indicates the effects in descriptive terms. The story is simple and logical, and it has the essential elements of cause-effect. It is easy for an owner or arbitrator to understand.

Phase 2

The information for this phase follows the same pattern as for Phase 1. The primary difference is that Phase 2 is more easily divided into three subphases, even though all the work is included in Phase 2 because the factors affecting efficiency, as shown on Table 7.5, are the same. Subphases are defined because it is easier to convey how the conditions on the top two floors affected the work more severely than the conditions on the lower floors. The most complicated courtroom work occurred on the top two floors.

- *Phase 2a.* Table 7.7 describes Phase 2a, which encompassed from January 14, Year 2 to February 11, Year 2. Little is made of the fact that the framing contractor worked for four or five months without a signed contract and appeared to have been given preferential treatment. This fact could not be substantiated, and such allegations would

Table 7.6 Relevant Facts for Phase 1

Key Events	Steel erection subcontractor does not complete steel until December, Year 2. Framing/drywall subcontractor does not sign contract with the general contractor. Framing/drywall subcontractor mans job, but with few workers. Project schedule slips weekly. Pressure mounts to make substantial progress.
Nature of Work	All the work by framing and electrical subcontractors done in the existing structure. Electrical sub increases manpower faster than would normally be expected in order to meet the original schedule. Much of the work by framing sub involves layout of wall track. Little or no framing is done. Electrical work in existing building is limited.
Causes of Labor Inefficiency (Electrical Sub)	The work is piecemeal. There is limited production work. Systems that integrate or occupy the existing building and infield cannot be done. Limited work in individual rooms is done. The work is done out of sequence.

potentially become personal and would deflect attention away from the core issues. During Phase 2a, the work performed was primarily on the top two floors. Table 7.7 explains how being "short tethered" to the framing operation caused the electrical contractor to use inefficient methods.

- *Phase 2b.* The key events, nature of the work, and causes of inefficiency are summarized in Table 7.8. This subphase went from February 12, Year 2 to March 25, Year 2. The work was largely on the second and third floors where the work was more efficient than on the fourth and fifth floors.
- *Phase 2c.* In this phase that ran from March 26, Year 2 to April 29, Year 2, the electrical contractor began to run out of work. Workers were sent back to the less-efficient fifth floor to complete more work, but overall, the finish electric work could not be performed because the millwork subcontractor had not completed work on any single floor. Table 7.9 summarizes this phase.

Figure 7.9 summarizes the relationship between labor inefficiencies and the floor on which the work was performed.

Table 7.7 Relevant Facts for Phase 2a

Key Events	Recovery schedule implemented. The order of completing the floors is reversed. Infield area available in early January. Framing subcontractor signs contract with general contractor in February.
Nature of Work	Framing subcontractor begins both framing and drywall on the 5th and 4th floors. Drywall closely follows framing forcing electrical contractor to be "short tethered" to framing contractor. Framing contractor sequence is walls, then ceilings, instead of room to room. Electrical contractor uses inefficient sequence because walls and ceilings not done together. Revisits to a room are required. Work on 4th and 5th floors is very disruptive.
Causes of Labor Inefficiency (Electrical Sub)	Electrical contractor must accelerate its work and perform rough-in between framing and drywall. Work must be done in the order and pace established by framing contractor. Work is done in a very inefficient sequence because of the lack of framing and ceiling coordination and drywall leaves no room to maneuver.

Phase 3

This phase is presented in Table 7.10. Most of this phase was spent in performing finish electric work when available. The work was severely strained because the millwork subcontractor did not prosecute the work aggressively or in the manner laid out in the CPM schedule. Figure 7.10 shows the worker days per week for the millwork contractor by floor. The figure shows that work progressed on all floors simultaneously, and that the work on all floors was completed at essentially the same time. This simple graphic effectively drove home the point that the general contractor allowed the work to be randomly performed to the detriment of the electrical contractor. Project photographs show clearly how completed areas were used by other specialty contractors to store materials, thus denying the electrical contractor access. These photographs are not shown in this chapter.

Table 7.8 Relevant Facts for Phase 2b

Key Events	Framing effort by framing contractor shifts from the 4th and 5th floors to the 2nd and 3rd floors. When rough-in work is complete on the 4th and 5th floors, electrical contractor sends electricians to 2nd and 3rd floors. Drywall work shifts to 2nd and 3rd floors when work on 4th and 5th floors is complete.
Nature of Work	Drywall lags sufficiently behind framing to allow electrical contractor to work more efficiently. Wall and ceiling coordination less of a problem. Overall, 2nd and 3rd floors less disruptive than 4th and 5th floors because electrical contractor has a short tether to framing contractor.
Causes of Labor Inefficiency (Electrical Sub)	Lack of wall and ceiling coordination means that electrical contractor must make multiple visits to many of the rooms. The sequence of work is inefficient. Efficiency improves when not working on the 4th and 5th floors.

Table 7.9 Relevant Facts for Phase 2c

Key Events	Mixed signals regarding the schedule. Substantial completion expected by the end of July. Drywall/plaster to finish on time. Concerns over coordination of some trades. Concerns over submittals and material delays. Concerns over letting things slip. Start of terrazzo delayed.
Nature of Work	Having completed work on 3rd floor, electrical contractor sends electricians back to the 5th floor. By the end of March, electrical contractor begins to run out of work because the rough-in is being completed, but millwork contractor has not completed millwork on any floor. By April 29, rough-in is substantially complete on all but the 1st floor.
Causes of Labor Inefficiency (Electrical Sub)	Electricians return to the inefficient environment on the 5th floor. On the other floors, there is less and less work available.

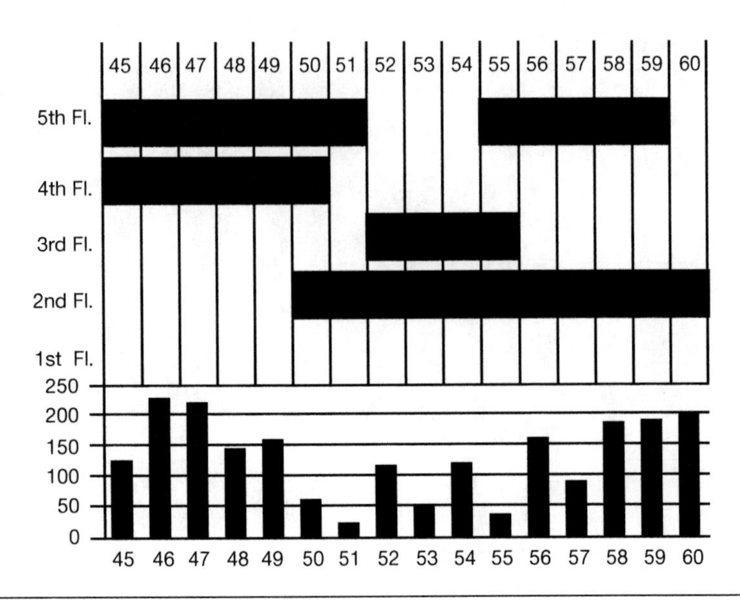

Figure 7.9 Relationship between inefficiency and location during Phase 2.

Table 7.10 Relevant Facts for Phase 3

Key Events	Early July, general contractor warns all specialty contractors of not meeting schedule. Terrazzo falls behind schedule. CPM schedule updated infrequently. Schedule continues to slip.
Nature of Work	Finish work stretches on and on. Millwork contractor's work never seems to end and is in progress on all floors. Electrical contractor must wait for millwork contractor to finish. Electrical contractor reduces work hours and responds when work becomes available.

Phase 4

During Phase 4, intense acceleration occurred. This phase is outlined in Table 7.11. Most of the inefficient work hours on the project were incurred during this phase.

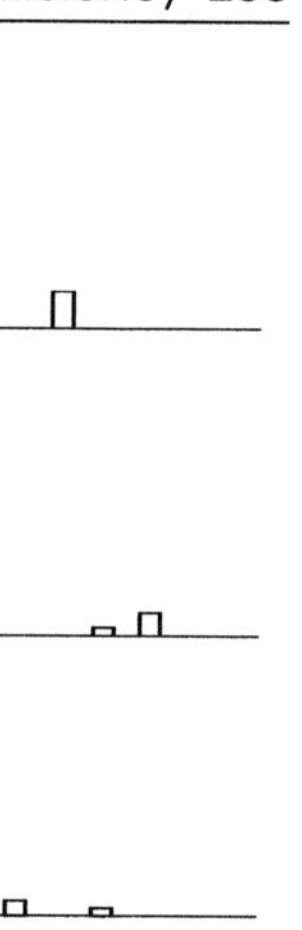
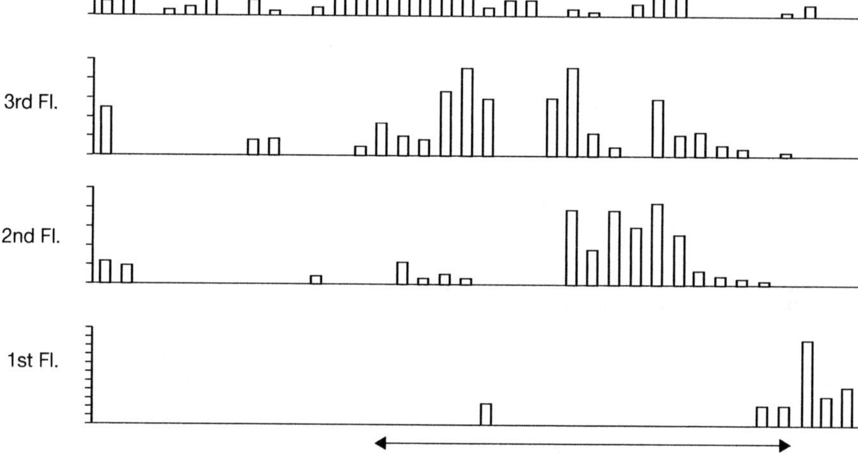

Figure 7.10 Worker days per week for millwork contractors.

Table 7.11 Relevant Facts for Phase 4

Key Events	Pressure to achieve Oct. 10 substantial completion date. Completion schedule slips. Substantial completion achieved in November.
Nature of Work	Electrical contractor installs fixtures and devices, completes systems such as fire alarm and courtroom sound, and works on punchlist items. Most of electrical contractor's work requires all other work to have been completed. To achieve substantial completion, electrical contractor increases his/her workforce to a high of 23.
Causes of Labor Inefficiency (Electrical Sub)	Intense acceleration. Overcrowding and stacking of trades. Disorderly process with little time to plan the work.

Table 7.12 Work Hour Inefficiency Summary

Phase	Total Work Hours	Inefficient Work Hours
1	3,517	1,664
2	7,617	2,153
3	6,710	0
4	4,090	2,764
Total	**21,934**	**6,581**

Finally, Table 7.12 summarizes the labor inefficiencies on the total project. It is estimated that the electrical contractor incurred a loss of 6,581 work hours, or a loss of efficiency of 43%.

Synopsis of Presentation Exhibits

The exhibits summarized in this chapter were used in an actual arbitration against the general contractor. The exhibits and presentation followed a number of important principles. These are briefly explained in the following.

The presentation tells a simple "Dick and Jane" story. The facts are condensed only to those that are relevant. Simplicity is also shown in how the events are related to a very few causes of lost efficiency, primarily piecemeal and out-of-sequence work, inefficient methods, and acceleration that resulted in congestion and stacking of trades. The presentation exhibits contain no complaining or personal assertions about the general contractor. It is both professional and factual. An important source of the factual information was the minutes of the progress meetings. Therefore, the facts were not in dispute.

If there is no showing of a cause-effect relationship, then the likelihood of recovery is minimal. The exhibits show this relationship by detailing key factual events and showing how these events affected the contractor's work. The electrical contractor's work was forced to be out of sequence and piecemeal. Inefficient methods were used, and in the last phase, stacking of trades and overmanning were significant factors. Notice how numerous detailed facts have now been related to a few causes of inefficiency. Owners and arbitrators can readily understand how these conditions affect efficiency. If not, experts can explain the effects and support their conclusions with data from literature and actual projects.

Lastly, the exhibits are simple and straightforward. They do not contain excessive information and are easy to comprehend. A copy of the exhibits can be provided to the owner or arbitrator, allowing notes to be taken thereon. They can be reviewed at a later time.

CONCLUSIONS

When negotiating compensation for loss of labor efficiency, the recommendations provided in this chapter should enhance the likelihood of recovering the additional labor costs. When presenting a claim, the emphasis should be on educating the owner or general contractor as to the causes and consequences of labor inefficiencies and on establishing cause-effect relationships to substantiate the labor overrun.

The presentation should tell a simple story limited to the relevant facts. The facts should be properly documented. No complaints or personal assertions about the general contractor or the owner should be included, as the objective should always be to generate a professional and factual case.

8

FIELD INCENTIVE SYSTEMS

Dr. Russell Walters, *Iowa State University*
Dr. James Rowings, *Peter Kiewit and Sons*
Dr. Mark Federle, *Marquette University*
Michael McArtor, *Iowa State University*

INTRODUCTION

The law of supply and demand drives the exchange of goods in this country. This law also governs the wages and salaries individuals can be paid for their services. In the majority of business enterprises, companies are free to base employee pay scales on individual merit. Companies that are obligated to work with employees that belong to a collective bargaining system will face restrictions including (1) the need for referrals from the local union office for increases in workforce, (2) reductions in workforce that must begin with the last people hired, and (3) a minimum pay scale that is determined by the collective bargaining contract.

The unions bargain collectively to secure better wages and benefits for the whole rather than the few. Pay scales are determined by perceived responsibility. Typical classifications include (1) apprentices—journeymen in training, (2) journeymen—typical craft workers, (3) foremen—supervisors of journeymen, and (4) general foremen—supervisors of foremen.

This system creates a dilemma: how does a company reward an employee for outstanding service when the company must pay wages according to the existing union pay scale? The curse and the gift of the union labor system is that each worker of equal standing in the union can expect the same wage on every job. This gives the union greater bargaining power when determining which workers will be sent to each job. The bargaining power is further augmented when the supply of skilled labor is down and the demand is up, as seen in several occasions within the construction activity. Thus, a company cannot withhold benefits or compensation from union employees unless it breaks its union agreements or risks having its labor needs go unfulfilled by the employees, who can expect to receive their minimum wage on any other project with any other company.

In simple terms, this is a discussion on competition—the competition between companies to secure the best field labor employees for their projects. The construction industry depends upon competent, well-trained field labor. Given the current legal climate, which is geared toward holding companies liable for their workers' actions, the increasing complexity of projects, and the greater speed at which projects are now completed, field laborers' abilities directly affect the success or failure of projects. This situation creates a very important need in the industry: to provide incentives to outstanding craft workers who are productive, safety-conscious, and loyal.

INCENTIVES ACROSS INDUSTRY BOUNDARIES

One similar use of incentive-based pay occurs in the commercial trucking industry. The skill levels of commercial drivers vary widely, and there are limited methods available to help employers identify superior drivers. Also, the demand for skilled drivers exceeds the supply. Contractors find themselves in a strikingly similar position. The commercial trucking industry is reacting to these circumstances by implementing merit-based and incentive-based pay systems that help companies recruit and retain outstanding personnel and create a sense of ownership among employees. A good example is the program instituted by USA Motor Express, based in Florence, Alabama (Huff 2001). The company created a standards-based incentive program called Team USA for its commercial drivers. To qualify for the plan, drivers must keep out-of-route miles under 10%. Their miles per gallon and fuel cost must match the average of the top 50% of drivers, and they must keep idle time under 45% of time worked. Also, they must have a clean driving record. The incentive is a \$1,000 savings bond each fiscal quarter and an additional \$1,000 savings bond to drivers that qualify for four straight quarters. This program has two goals: reward the company's best drivers with a \$1,000 incentive, and try to retain employees by providing an additional incentive if they have four straight qualifying quarters. By following this program, the company has increased its number of trucks by 50% or more each year. The company has been profitable since this operation began.

Employees Express trucking lines of Houma, Louisiana (Kelley 2002), created a slightly different program. This company sets aside a portion of its profits, usually around 33%, for a year-end incentive payment to employees. The money is allocated to the employees following a point system based on income and number of years served. For every year an employee has worked at the company, he or she gets 5 points, and for every $1,000 earned in salary, he or she gets 1 point. Using the point totals, the company calculates how much money to deposit into each employee's retirement account. This plan encourages driver retention by rewarding seniority and by using a vesting system. Employees do not gain full access to their retirement account until they have been with the company for five years. If they leave the company after only two years, they will control only 40% of their account. The effect of the program can be seen in Employees Express's recruiting: there is a waiting list of drivers ready to join the company.

Companies like Quality Distribution (Nicholas 2001) give pay increases based upon merit alone. This company has one pay rate for all of its drivers, and bonuses are based on factors such as on-time performance and miles traveled without an accident or ticket. Another example of a successful incentive system comes from a technical staffing service formerly called Mid-States and now known as ENTEGEE (Case 1994). This example is appropriate because the company provides temporary help to other companies for a fixed fee, plus incentives provided by Mid-States. This service mirrors the labor staffing practices used by unions in that it is of finite duration and the employees serve two entities at once: the company to which they have been temporarily assigned and the umbrella company to which they belong. Mid-States rewards employees based on a "Bucket Plan" in which six imaginary buckets are filled with cash from company profits. When one of the buckets is full, the company makes a payout. The size of the buckets is determined by the company's pretax profit goal for that year.

The remarkable element of this plan is not its design or implementation, but its results. When a study of the plan was completed, the company found that it was paying out an average of 15.6% of employees' base pay per year. In conjunction with this, it discovered that the average base pay of all Mid-States' employees increased only 1% per year during the seven-year study period. Thus, the incentive plan largely replaced annual base pay increases as a primary employee incentive. The success of these programs, which all involve making payments from company profits, proves that incentive-based and merit-based plans are effective methods of accomplishing company goals.

An industry not discussed above is manufacturing. Lincoln Electric Company is an excellent example of a manufacturer with a successful and longstanding incentive program (Dawson 1999). In 1941, with the outbreak of World War II, the demand for welding products skyrocketed. Lincoln Electric found itself in the enviable position of supplying fully half of the welding needs of the United States, and net sales soared from $13.6 million in 1940 to $24 million in 1941. With qualified workers made scarce by the need for able-bodied men and women in the war effort, Lincoln Electric found that it could train and retain personnel during and after the war by using its fledgling incentive program. The program set aside a large portion of their yearly profits. At its peak, the program allowed bonus payments of 60 to 100% of an employee's base pay. The result of this program was that Lincoln Electric workers produced at four times the rate of their nearest competitor. According to J. F. Lincoln, the company founder, the high productivity was "because of the fact that they [the employees] own the company and share in its profits" (Dawson 1999).

The success the company garnered with the incentive program gave Lincoln an opportunity to further clarify his philosophy on incentive systems. In 1946, Lincoln's first book, *Lincoln's Incentive System* (Lincoln 1946) contained the following passage:

> *Incentive management is more like a religious conversion. It is not a spur to the man to speed up; it is a philosophy of work. It is not a method of getting more work for lower wages; it is a plan for making industry and all its parts more useful to mankind.*

In the modern era, Lincoln Electric continued to refine its incentive compensation program. The program now involves a direct link between the profitability of the company and an employee's bonus, since in its earlier incarnation this relationship was usually arbitrary. The company awards the bonus if its financial objectives are met and the individual's performance warrants the bonus.

Lincoln Electric's success with its program illustrates two key points. First, higher productivity is achievable by giving employees a stake in the company's profits. Second, a properly constructed incentive program is not a gimmick or a quick fix to solve a company's problems, but rather a shift in management philosophy that requires confidence in the system, from the top executive to the apprentices of the company.

The discussion on incentives so far has not remarked upon the uniqueness of the construction industry. Certain industries have a reputation for using incentive systems. These industries include software development, commercial trucking, investments, and product sales. The construction industry commonly uses incentive systems for employees working at and above the level of project manager. The need for an incentive system in the construction industry is becoming more critical, and this need is particularly critical for field personnel where several studies project a shortfall in available labor in the foreseeable future. This will guarantee that competition for good employees will increase, and contractors will be required to find ways to ensure that their projects are adequately staffed. To satisfy this need, an incentive system will have to be tailored to the industry. This industry is unique in the sense that two companies, with equal legitimacy, can call themselves contractors even if they perform very different types of work. Because of this diversity, designing an incentive system that meets the needs of all contractors is challenging. This challenge will be addressed in more detail in the sections dealing with the design of the incentive system, but it is important to recognize the uniqueness of this industry.

CHARACTERISTICS OF CHANGE

Companies like to receive assurances, prior to the start of a new program, that the program will ultimately be successful. The decision to design and use an incentive system can also be evaluated in the light of a few basic criteria to determine whether the system's implementation will be successful. In this case, "successful" does not necessarily mean that the program will produce the desired effect; rather, it suggests that management, the administration, and the employee-stakeholders accept the program from the beginning.

People evaluated by incentive programs often view them not as something positive but rather as another instance of "Big Brother is watching." This idea can create mistrust and fear among employees if they believe the program has been generated not to reward the outstanding but rather to seek out and punish the underachiever. Therefore, effective means of communication must be in place before the incentive program is begun. The employee-stakeholders must be included in the design process to the point that they feel comfortable with the evaluation criteria. This inclusion can take the form of an employee forum or an informal lunch with a few members of the workforce to allow them a sneak peek at the program before implementation. This will not only head off possible problematic criteria but will also give employees confidence that the evaluation criteria are really in place to reward and not to punish.

With respect to each of the key players in the incentive program (management, administration, and employee-stakeholders), a few key characteristics and responsibilities should be in place during or before design of the plan.

The diagram shown in Figure 8.1 describes the relationship between these three parties. Management is responsible for generating the performance goals and the criteria that will be used to evaluate employees' attainment of these goals. The selection process for the goals will be discussed later on in this chapter, but managers must have a prior grasp of the goals they want to accomplish with the program and the criteria the company is capable of measuring. For example, while it might be

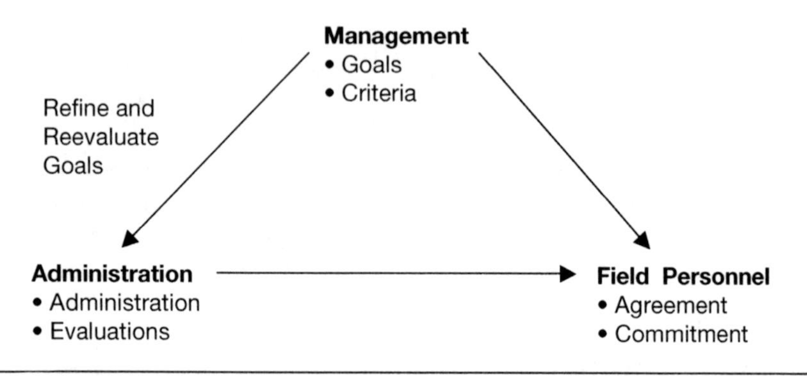

Figure 8.1 Requirements and expected results for the incentive program.

desirable for a company to measure installation speed, if the company installs thousands of items a year, this might not be a feasible measurement to track.

Some form of administration or clerical staff will be required to run the program. If a company has a minimal administration staff, then it probably does not have the ability to implement a complex incentive plan. A company should verify that it has sufficient resources to ensure the program can be implemented once it has been designed.

Finally, employee-stakeholders have to exhibit two basic traits to justify the incentive program: a willingness to accept responsibility for the company's goals and a commitment to a new method of incentive compensation. Historically, companies sometimes succeed without obtaining this support prior to implementing an incentive program, but with today's unionized workforce, it is better to be safe than sorry. In most programs, the minimum an employee can make is the agreed-upon base salary, so employees have nothing to lose and everything to gain from accepting the program. A company should emphasize this during the opening discussions concerning the program.

DESIGNING AN INCENTIVE SYSTEM

The first question to ask in designing a new incentive system is this: What is the system supposed to do? It would seem unfair to ask any company owner, board of directors, or human resource manager to implement an incentive system if some benefit was not expected. After all, the incentive system's operation will include giving a significant portion of the company's profits to the employees, in addition to their regular wages. Most would expect benefits such as the following:

- Increased profits
- Increased safety awareness (leading to fewer accidents)

- Increased adherence to company policies
- Increased reliability and better performance of schedules (including accurate work projections and increased work rates)
- Increased loyalty to the company (leading to decreased turnover rates)

It should be noted that this particular incentive system's design is not how it should be done, but rather how it could be done. A company must evaluate its own needs and set its own goals when designing its incentive system. The system that we are discussing in this chapter is seeking to satisfy each of the goals discussed above, but dilution of substance is a concern in incentive systems. If a company has specific concerns that need to be addressed, an incentive system that targets those concerns will produce a greater impact than a more general plan in which the targeted concern is mixed in with secondary concerns.

Robert Sibson (1990) recommends that a company's goals be as objective as possible to avoid distrust or confusion among the employees. General goals such as "improve morale" or "improve the quality of life" lack substance and seem to have hidden agendas. Defined goals such as "reduce employee turnover to 10%" or "increase profits by 3%" are more effective and will help the company to assess the worth and progress of the incentive system. For the purpose of this design, the company goals will encompass all of the goals listed above. This will also serve as a model from which companies can build their systems.

Looking at the basics of a compensation program, Sibson has generated eight ideals or objectives upon which the incentive program for contractors can be constructed.

Objective 1: Serve All Stakeholders

An incentive system must always seek to create a balance between the owner-stakeholders and the employee-stakeholders. In the case of this incentive system, the employee-stakeholders are the field labor, and the owner-stakeholders are those looking to accomplish the goals set forth by the company. By their very nature, incentive plans are an exchange program. The owner-stakeholders give a portion of the profits to the employee-stakeholders in order to accomplish their goals. The employee-stakeholders provide the "horsepower" to drive the company to the goals in exchange for additional compensation.

The owner-stakeholders should view this exchange as an investment in their company goals. A company might be able to function and make a profit without having clear goals. However, a company with an incentive plan seeks to make the employee-stakeholders accept responsibility for accomplishing the company goals by including them in the rewards.

The goals of most union members are to have steady employment, make a good wage, and avoid injury. Many members and companies also desire to produce quality workmanship, but this is more a function rather than a goal. However, fieldworkers can accomplish their goals without involving themselves in the company's success or failure. Few people would accept responsibility for a set of directives for which they felt they were not compensated or over which they had little control. The incentive system places the goals of the company before the employees, asking them to take responsibility for these goals and compensating them for doing so (Hessen 2000). As a result, this system should balance the responsibility that employee-stakeholders accept with the incentive the owner-stakeholders supply.

Objective 2: Keep It Simple

Being able to effectively communicate the incentive plan to employees will be of paramount importance once implementation begins. How can one expect the employees to focus on the goals of the incentive plan if they do not understand how their individual performances contribute to the company goals and thus to their reward? Keeping the plan simple will make it possible to discuss the plan with the employee-stakeholders and eliminate the suspicion that there are hidden agendas or unfair provisions in the agreement.

Second, a simple plan will allow for simple administration. Incentive plans will inevitably create additional paperwork and administrative tasks necessary to run the program effectively. If the program is straightforward and clear, it will require less work and less training time from those responsible for its operation. An incentive program does not have to be a burden to administer, and many companies will find that their plans can be attached to their existing employee evaluation programs. In effect, a good incentive program can be measured by both its results and its simplicity.

Objective 3: Start by Identifying Company Needs

Incentive plans are created to solve problems within a company. Given that, a company must know what problems it wishes to solve in order to properly institute an incentive plan. Identifying the problems to be overcome by the incentive plan should be accomplished via the same process a company uses to identify any problem. Creating an incentive plan to solve a problem that does not exist or the wrong problem will indicate that the incentive plan failed, when in fact it never had the chance to succeed.

Sibson (1990) recommends a three-step process to accomplish this objective. First, the company must identify a clear and compelling need. Without a clear

Table 8.1 Sample Profit Distribution

Division	Percentage
Executive	25
Management	25
Estimating	25
Field Labor	25

need, why not just keep the profit within the company? Second, the need and the solution (the incentive plan) must be clearly connected, or, in other words, the incentive plan must solve the problem. The conclusion that the incentive program *might* solve the identified problem is unlikely to convince the owner-stakeholders to commit a significant percentage of their profits to the program. Finally, the value of the program must be greater than the cost. A value-to-cost ratio of 4 to 1 is suggested, since cost estimates can often be vague. A company must not forget the costs associated with the administration of the program as well as the final payout every assessment period.

Objective 4: Group Employees Properly

When deciding how to allocate the profit to the various employee-stakeholders, it is customary to create a distribution of the net profit to each group. A sample profit distribution scheme is shown in Table 8.1.

This distribution structure is of greater concern to a company that already has an incentive or profit-sharing program, but it is still a consideration for a company dealing with only field employees. When a company organizes its program, the distribution of profit among the various job types (project manager, journeymen, and clerical staff) must be done to provide the proper level of incentive to each group. While there is no canned set of numbers that can give a company the desired outcome, the incentive system proposed in this chapter is based upon typical numbers in the industry and thus can serve as a starting point. Some amount of trial and error and what-if scenarios are necessary during the development of any program to get the mix just right.

Within the subject of field personnel, a company can choose whether to recognize and utilize the union designations (apprentice, journeyman, foreman, etc.) to determine the groups. Nonunion contractors may choose to group employees by years of service or to treat all the same. The decision is based largely upon the need that the program is trying to solve. If retention of senior personnel is a problem, then it would behoove a company to recognize these differences.

Objective 5: Have a Proper Process for Developing Compensation Programs

It goes without saying that companies have different challenges and problems. These differences will require any incentive program to be customized. As stated previously, an incentive program's primary purpose is to solve a problem, so that problem should always be the starting point for any program. Many companies find utilization of a checklist a good method to develop a program. Sibson (1990) recommends the following process:

1. Identify needs, problems, or opportunities, focusing on such methods as the following:
   - Data analysis
   - Reporting issues
   - Discussions with project managers
   - Discussions with field personnel
   - Discussions with other employees
   - Personnel audits
2. Develop objectives, such as the following:
   - Set specific goals
   - Consider impact on other human resources programs (such as existing profit-sharing programs or bonus structures)
   - Consider program schedule (implementation, frequency of payouts, etc.)
   - Consider resources available (current financial standing, profit projections, etc.)
3. Conceptualize the answer early, considering the following:
   - Company characteristics
   - Competitive practices
   - Company culture
   - Employee reactions
4. Consider testing and specific program design, which may involve the following:
   - Legal, tax, and accounting considerations (how the program affects financial factors such as cash flow and investments
   - What-if scenarios and modeling (testing the program for loopholes, inconsistent results, and unintended rewards or punishment)

- Evaluation against company plans and forecasts (an opportunity to recheck the program's value and costs against the company's future earnings)

5. Implement the program
6. Evaluate the effectiveness of the program
7. Review the program

Objective 6: Design Program to Reflect Company Culture

A program tailored to produce an incentive for field personnel will face unique challenges. There are long-standing and highly ingrained preconceived notions between management and labor that could interfere with successful program implementation. Labor might be inherently distrustful of any new initiative proposed by management, and managers may not be able to accept the idea that labor is being compensated beyond what they perceive as a healthy wage.

Unfortunately, there are no simple solutions to these obstacles. The best resource is open communication about problems that the plan is attempting to correct. Owner-stakeholders will need to accept the fact that the incentive plan will make the company stronger and will assist them in creating an elite workforce. Employee-stakeholders will also have to be assured that the goals of the plan are in their best interests, without undesirable side effects.

Overall, the program must be compatible with the existing culture of the company. The institution of a rigorous and intensive program will not work for a company that has a casual, informal management style. When developing the incentive program, the company must ask if the program is similar to management initiatives that have been produced in the past, and if employee-stakeholders would accept this style or amount of observation.

Objective 7: Install a Formal Program

The new incentive program should be formalized and include standards and application rules that are discussed with and understood by both the owner-stakeholders and the employee-stakeholders. Stakeholders will trust a formal program, which in turn will be more effective because the rules of the program, when followed properly, produce the desired reward and effect. A program that seems to be based upon arbitrary factors will give the employees a "pennies from heaven" perspective on the program: When they are rewarded, they will not believe it is the result of their own outstanding performance but rather the work of some unseen benefactor who drops the reward into their laps.

A formal program does not have to be an elaborate program. The benefit of a formal program is that it produces trust on the part of employees; an elaborate or complex program that employees do not understand will most likely be a program they do not trust.

Objective 8: Be Skilled in the Art of Managing Change

Nobody is perfect. This simple, yet true, statement applies to the business of creating an incentive system. The managers who operate and apply the incentive system must be skilled in the art of managing change. The program will most likely need adjusting after implementation has begun. Changing the program "midstream" might produce adverse reactions on the part of the employee-stakeholders who believe the changes are a sign of upper management's changing priorities. As it is typical in the construction industry, better communication can solve this problem. If a problem has slipped past the early tests and what-if scenarios, it is better to change the program to ensure that it will produce the desired effect rather than allow it to continue working toward the wrong goal. Communicating with the employee-stakeholders and keeping them involved in the implementation process can alleviate fears and confusion during the fine-tuning phase. After all, an employee rarely begrudges a company its goals. Goals such as "increase profit" and "improve safety" can be translated as "ensure there is a company for employees to work for in the future" and "ensure employees go home at the end of the day." So if changes must be made to make certain the program accomplishes its goals, then letting the employees in on the reason for the changes can make them more receptive to them.

BASICS TO SPECIFICS: DESIGNING AN INCENTIVE SYSTEM FOR FIELD LABOR

In designing an incentive system for field personnel, one of our driving factors was to have the finished product serve as a model that companies could customize for their own use. For ease of discussion, the company for this design example will be called Sparks, Inc. As a starting point, the design goals for this incentive system are as follows:

- Increase profitability (by project) by 10%
- Decrease experience modification rate (EMR) by 0.1
- Increase schedule conformance to have all projects finished within their estimated worker hours
- Increase marketing efforts by the field personnel (target: one referral per employee per year)
- Increase retention of experienced, certified, and well-trained personnel

- Increase conformance to company policies and job tasks (for instance, proper attire, customer relations)

In the forthcoming design, each of these goals will be addressed in a category, or several categories, within the overall incentive system. Sibson's eight basic objectives encompass the issues that are pertinent to this design, but for the sake of clarity, the salient points will be touched on again.

For this case, Sparks, Inc. has decided to address several issues. It wishes to increase profit, lower EMR, improve schedule performance, increase market share, increase retention of top employees, and increase conformance to company policies and standards. This hefty list of goals common to many companies is as specific as possible and includes particular target goals to allow easier assessment of the effectiveness of the program. Sparks, Inc., has decided that it will include all people who worked on at least one of its projects during the previous year and that it will pay out incentives annually.

One of the most difficult determinations a company must make when designing an incentive system is how much profit needs to go to the program. Too much profit committed means the company has less money to invest. Too little profit committed and the program will not provide sufficient reinforcement of the evaluated behavior. There is no magic number that can be suggested, but as a practical response, the amount of money to be invested in an incentive program should be determined in the same manner that a company uses before undertaking any investment. If the future benefits and earnings of the program justify the up-front development and yearly costs, then the program is a good investment.

In most programs, the minimum payment of 1% is a cutoff. If there is not enough money in the program to pay at least 1% to each employee-stakeholder, no payments are made. This is to ensure that employees do not receive a dollar amount so low that it discourages them rather than encouraging them to participate in the plan. Each plan administrator should designate the minimum payment for the plan based upon how rigorous the plan is. For example, a plan that requires little effort on the part of an employee could have a very low percentage set as the minimum incentive. In contrast, a plan with very rigorous requirements must have a higher minimum payment to avoid angering the employees.

Companies may also wish to establish a maximum program payout to ensure that there is some measure of predictability in costs. While this number must be set and communicated to the employees, no particular percentage or number can be recommended. Each company will need to evaluate its needs and decide what, if any, maximum percentage should be set. The incentive system designed in this chapter has leveling built into the formulas to prevent any one employee from receiving a disproportionately high bonus, but exceptions can and will occur.

It should be noted that the information presented on typical incentive payouts included two industries (financial services and diversified services) that have a history of high incentive-based earnings; thus, one could expect that a respectable payout for the construction industry would fall between the minimum and typical percentages. This would correspond with a percentage payout close to 5% to 6%.

Now that Sparks, Inc. has chosen its goals and the amount it is willing to spend, it must select a methodology to link job performance that helps the company accomplish its goals to the appropriate monetary reward. This is accomplished using employee evaluations.

In industry today, approximately one-third of companies with incentive compensation plans utilize employee evaluations (Parker et al. 2000). In addition to evaluations, companies can also use market pricing and market classification to determine the level of fair compensation. However, in the case of field labor, this is not an option. Wages for all field electricians are set prior to their employment by a company. This does not allow market factors to determine compensation. The response to this problem is to use employee evaluations.

The methodology used in employee evaluations has changed little in 50 years. It requires the evaluator to measure an employee's job duties against a predetermined benchmark in order to determine relative job worth.

In the interest of creating a quantifiable scale, the employee's "relative job worth" can be calculated using a point value system. Simply stated, a point value system is one in which criteria are evaluated on a separate scale for each factor and weighted points are assigned based upon the relative worth of the criteria. In some cases, the weight of a criterion might be easily determined, such as in the case of safety, where one injury is much better than three injuries, or it may be more arbitrary, such as assigning points based upon the appropriateness of one's attire. Among the advantages of a point system we can cite the generation of a clear record, the applicability of statistical methods to determine the effectiveness of the program, and the simplicity of administering the program.

Matching the company goals to applicable job duties with appropriate point criteria is the next challenge in the design of Sparks, Inc.'s incentive system. There will be four general categories: profitability, safety, job tasks, and marketing.

Profitability

One of the most difficult yet critical factors to assess is profitability, especially as it relates to field labor. Invariably, the question of productivity will arise as it applies to profitability. When most managers discuss the effectiveness or efficiency of their field personnel, they refer to their employees' productivity. The problem stems from the fact that productivity covers an unlimited range of possible activities—all of which have different impacts upon profitability. Asking a manager to keep track of the productivity of each employee to the level of detail required to establish an evaluation would be a monumental and costly documentation exercise.

Rather than undertaking this time-consuming challenge, one can assess profitability without determining each employee's productivity. Informing fieldworkers that a good portion of their yearly bonus is tied to the overall profitability of the projects they work on will induce them to work at their best-possible speed without requiring additional monitoring.

This action really has two purposes: It ties company profitability to individual project profitability, on which the fieldworkers have a direct impact, and it creates a team mentality among the workers by assessing profitability by project. If workers know that another worker is not pulling his or her weight in the field, and they know their bonuses will suffer because of it, they will "self-police" by either weeding out those that cannot handle their share of the work or assisting slower workers to bring them up to speed. Furthermore, workers who find they are falling behind due to a lack of ability in a certain skill area will be more likely to seek out specific training, even at their own cost, if they know that they will be reimbursed via the incentive program when their improved performance results in increased rewards.

To create a point system based on profitability, there must be a weighting factor that can determine the relative point value for this category. For this design, the point total is obtained by dividing the total project profit by the planned project profit and multiplying by the weighting factor. For instance, if the weighting factor is set at 1, then it is worth 1 point if the employee matches the profit goal of a particular project. The greater reward comes when this fraction returns a number greater than one, which occurs when total project profit exceeds the planned profit. This point deserves emphasis because Sparks, Inc. is attempting to increase profits.

The weighting factor is not as critical in comparisons between employees because they will each receive the same number of points for the same amount of improvement; thus, it makes little difference to them if the weighting factor is 1 or 5 as long as it is the same for all employees. This index value becomes important when it is compared to the other categories under evaluation.

If Sparks, Inc. sets a design target of about 25 points for an average employee, then it can check the importance of profitability against that number. Assuming a good crew can exceed profit expectations by about 25%, then the profit factor would be 1.25. If the weighting factor were set at 6, then profitability would be about one-third of the total points for an employee, which is acceptable. If a company wishes to emphasize profitability above all else, then this weighting factor could be set at a much higher number, such as 10 or even 20.

The question of whether to include change orders in determining overall profit is up to the company. Typically, the profitability of change orders can be measured in the same way as initial profit projections; thus, Sparks, Inc. has decided to include change orders in its formula.

In order to account for the fact that an employee could work on several jobs in one year, each project's profitability can be determined and then points assigned based upon how long the employee worked on a given project. The equation can then be applied to each job and the point values totaled. Equation 8.1 shows the formula used to determine points due to profitability.

$$\text{Points} = \left(\frac{\text{Total project profit}}{\text{Planned project profit}}\right)\left(\frac{\text{Hours assigned to project}}{\text{Total working hours in plan cycle}}\right) \quad (8.1)$$

Working hand in hand with profit is the timeliness of project completion. One of the company's goals was to increase schedule adherence so no project finished with more worker hours than it was originally allocated. Given that change orders can lengthen the schedule for a project, Sparks, Inc. will evaluate employees only based upon contract work. Equation 8.2 will be used to generate the points earned as a result of timeliness of project completion. For this evaluation, the points will also be determined and weighted according to how long an employee served on a particular project.

$$\text{Points} = \left(\frac{\text{Planned working days}}{\text{Actual working days}}\right)\left(\frac{\text{Hours assigned to project}}{\text{Total working hours in plan cycle}}\right) \quad (8.2)$$

This scale should be applicable to projects of different lengths. Most companies have projects with a wide variety of schedules, ranging from a few hours to a few years. There will be occasions when projects extend over several incentive periods. This situation can be handled in a few ways. The company can delay incentive payments until the project is complete. This may require the company to properly weight this type of project to compensate the employee for having to wait for the bonus until the later date. The company may also take a "snapshot" of the project's status when the incentive period ends. This will allow the company to be up-to-date on the project and provide more immediate incentives to the employees. Finally, instead of running the program on a yearly or other time-based assessment period, the company could run the program on a project-by-project basis. This would allow the program to be up-to-date on every project and would give the employees an immediate incentive payment when the project is complete.

Each method has its advantages and disadvantages. The first method requires a minimum of administration, while the next two, by providing more immediate payments, do a superior job of tying performance to incentives. The second option has the obvious problem that if a project has a strong start but a poor finish, then the company might have to withhold further incentive payments to cover the premature payments on a project that went bad. This issue and others related to tying performance to the incentive program are further explored later in this chapter when we discuss how to customize this incentive program.

Safety

The safety category has been created and tailored to produce two effects. First, the category will reward employees for attending all required safety meetings. This will encourage attendance at the meetings and, in turn, will help the company communicate with employees about its safety program. Each company can decide how to weight this category. In this case, Sparks, Inc. has four company-wide meetings per year that all employees must attend, as well as daily "toolbox talks" and other safety meetings. Each meeting attended is worth 2 points.

The company also wishes to reward employees who go beyond the call of duty and attend additional safety classes not affiliated with the company. For example, an employee who attends a certified 10-hour OSHA or CPR course will receive points. For each approved outside course attended, the employee is rewarded with 2 points.

To encourage on-site safety conformance, Sparks, Inc. will penalize each employee 2 points for every lost-time accident the employee has in the plan cycle. The system is noncumulative, though, so an employee injured in a given plan cycle has the slate wiped clean in the next one. By evaluating these three aspects of employees' safety practices (and assigning points to them), Sparks, Inc. seeks not only to lower company injury rates, and thus insurance rates, but also to foster initiative and communication among the workforce.

Job Tasks

Each employee, in addition to his or her labor output, is usually required to perform some other administrative or documentary functions. At the journeyman level, this might be reporting work output or keeping track of tools and materials. At the foreman level, this could also include timekeeping and reports to superiors. Superintendents could be expected to perform additional timekeeping and management-level tasks, such as scheduling and estimating as well as manpower loading.

Employees' dedication to these tasks is instrumental to the proper operation of any company. Their ability to accurately report their progress and status has ramifications not only for the project at hand but also for future estimating and planning efforts. In order to encourage this dedication, Sparks, Inc. awards points for this category. For each of these tasks the employee completes satisfactorily, the employee receives 3 points. To keep this category properly balanced, Sparks, Inc.

must decide which tasks are required for each employee classification. Table 8.2 will be used for this purpose.

Notice that as the hierarchy increases, so does the number of required tasks. As a result, the parties accepting the greatest amount of responsibility receive the highest potential reward. This category may be able to reinforce other new programs, such as best practices or benchmarking plans. The company can make a job task that corresponds with a critical function of one of these other programs to ensure that the program is given the proper consideration by the employees. If a particular program is of extreme importance, the point value for its job tasks can be increased.

Marketing

In the contracting business, each employee is a salesperson to a greater or lesser degree. A field employee's interaction with the community at large is often overlooked as a viable and vital resource for the growth of business. The marketing category seeks to tap into this resource by awarding points to field employees who identify sales opportunities for the company.

If an employee identifies a reasonable opportunity for the company, then the employee receives 1 point. If the opportunity results in an actual work award, then the employee gains an additional 2 points. Furthermore, if an employee identifies a reasonable opportunity that is unpublicized, meaning it has not yet been submitted to the public record or for other contractors, then the employee receives 2 points. If the unpublicized opportunity results in an award, the employee earns an additional 3 points.

Rewards are set high in this category to increase field employees' initiative and awareness of the company's marketing. For example, if an employee submits one unpublicized opportunity and the company is successful in securing the work, then the employee receives 5 points. If Sparks, Inc.'s approximate total point-per-employee estimate is around 25, then a 5-point increase represents approximately 16% increase in reward.

Discretionary Categories

Companies might choose to add additional categories to encourage adherence to company policies, company standards for customer care, or other discretionary items. Sparks, Inc. is a service-oriented company that requires adherence to a dress code and employee customer service policy. Companies that have other concerns can tailor these categories to fit their needs.

Using this category, the company can also create the incentive for skilled workers to stay with the firm by rewarding employees for their years of service

Table 8.2 Job Task Requirements by Job Title

Status	Task
Apprentice/Journeyman	Report work output properly and in a timely manner
Foreman	Report work output properly and in a timely manner Prepare and submit accurately time keeping records
General Foreman/Superintendent	Prepare and submit accurately time keeping records Prepare monthly status reports accurately and submit in a timely manner Prepare manpower loading reports accurately and submit in a timely manner

and the number of certifications they hold. In order to keep this category simple and rewarding, for each certification (union- or trade-recognized) that the employee holds, he or she will receive 1 point. Such points will be awarded according to the plan cycle; thus, one certification held will be worth 1 point each plan cycle. This will also hold for any additional training courses that the company recognizes; thus they will also be worth 1 point.

For years of service, Sparks, Inc., has divided up the points scales as shown in Table 8.3. These years of service points will be awarded each plan cycle, but not cumulatively; thus employees with eight years of experience will receive 2 points each year until they reach ten years of service, after which they will receive 3 points each year.

Companies should recognize that there are pros and cons to awarding points based upon service years. On the pro side, the program encourages employees to work as many years as possible with an employer, which thus secures personnel for future projects. This is based upon the idea that keeping an employee already trained in a company's procedures and methods adds value to that company. Thus, the additional incentive payout is offset by the additional value the long-term employees bring to a business.

On the other hand, an employee who has worked for the company for a longer time is not necessarily a more valuable employee. It is extremely difficult to evaluate employees based upon their individual skill sets and to provide incentives based upon this information when there is no system in place to measure their individual skills. Most skill-level evaluations in the industry today are based upon subjective comments about a given employee, such as "he is a good hand" or "she knows her stuff." As a result, companies should use this category with caution and should understand its limitations. If a company is interested in awarding an incentive to employees who show basic loyalty to the company by working for it when possible, this category is applicable. If a company wishes to reward employees with the best skills, this category will probably not accomplish that objective.

Table 8.3 Years of Service Points Award

Length of Service	Points
0 to 2 Years or 0 to 4,000 Hours	0
2 to 5 Years or 4,000 to 10,000 Hours	1
5 to 10 years or 10,000 to 20,000 Hours	2
10 to 20 Years or 20,000 to 40,000 Hours	3
20+ Years or 40,000+ Hours	4

When administering the incentive program, companies must decide how to handle an employee who has been fired due to policy violation or an employee who has quit. If the employee had a major policy violation such as theft of company property and was subsequently fired, the incentive could be reduced or withheld altogether. However, an employee who quit for an acceptable reason such as geographic relocation should still receive an incentive based on the number of hours worked. No set of rules or points can replace judgment and common sense, so it will be left up to administrators and managers to decide how to handle these occurrences.

Additional categories can be created, almost at will, to meet the needs of the company. In this example, Sparks, Inc. is a service-oriented company that has an approved uniform (attire) code and has the customer complete a survey at the end of every project. The company can create two incentive categories to place a "spotlight" on these operations. Administrators must ensure, however, that the points dedicated to the new category are in balance with the rest of the program and appropriate to the amount of emphasis the company wishes to place on that issue.

With this design example, where the total points will be around 25 for an average employee, creating a single category worth 10 additional points would place a great value upon the new category and, in turn, devalue the other categories. Companies must review new categories and use judgment to ensure all categories meet the goals of the overall program.

A common complaint among contractors is that project performance suffers when a workforce has a high incidence of absenteeism. An incentive system provides a ready-made framework for controlling absenteeism. Most companies regularly track attendance and record unexcused absences. Creating a category that uses this existing information is fairly simple.

There are two possible approaches to a point scale system for this category: (1) deduct points for unexcused absences, and (2) set a maximum number of unexcused absences that will be tolerated before an employee is removed from the incentive program. In the latter example, a company might set a limit of five unexcused absences; employees who reach the sixth unexcused absence are notified

Table 8.4 Attire Point Scale

Status	Points
Always Worn	2
Often Worn	1
Sometimes Worn	0
Rarely Worn	-1
Never Worn	-2

that they are no longer eligible for the program. Companies will need to have a method in place for determining when an absence is excused or unexcused, but that question is best left to each company.

In a system in which points are deducted from an employee's total, each unexcused absence could be worth 1 point or a half point. The amount deducted should reflect the company's concern with absenteeism. For example, an employee with three unexcused absences could be penalized 3 points. If the average point total expected for a given employee is 25 points, then this would be a deduction of 12% from an absent employee's incentive payment. Obviously, employees with too many unexcused absences will see their point total drop toward zero or negative points, removing them from the plan. This category is provided here as a possibility, but it will not be included in the Sparks, Inc. example, for simplicity's sake. The other categories presented below will be incorporated into the design example to show how they, and other categories like them, can work.

Attire

At the end of the incentive plan cycle, an employee's direct supervisor is asked to make a judgment about each employee's adherence to the uniform code. An "Always Appropriately Attired" rating is worth 2 points, an "Often" rating is worth 1 point, a "Sometimes" is worth zero points, a "Rarely" subtracts 1 point, and a "Never" subtracts 2 points. This scale is shown in Table 8.4. Again, if this scale is published and the employees are aware this is an incentive item, then they will be able to either secure or reject additional compensation for themselves.

Client Feedback

Assuming that a service-oriented company has asked clients to complete a feedback form upon the completion of each job, then that form, too, can become part of the incentive program. Employees who receive high ratings can be rewarded for their attention to customer satisfaction. Granted, there can be indexing and customer apathy problems that render client feedback either unusable or biased, but if

Table 8.5 Client Feedback Point Scale

Client Feedback	Points
Very Positive ("4" Rating on the Survey)	2
Positive ("3" Rating on the Survey)	1
Neutral ("2" Rating on the Survey)	0
Negative ("1" Rating on the Survey)	-1
For Every Formal Complaint Received	-2

a company uses surveys, it trusts that the law of averages will keep this category fair. Thus, a company can create a category with a point scale as shown in Table 8.5.

These categories are subjective. They should be included in the incentive program only if they are based upon a trusted evaluation method that the company either currently uses or will implement.

CUSTOMIZATION OF THE INCENTIVE PROGRAM

The purpose of this section is to highlight the underlying issues related to tailoring an incentive program. As with most programs of this type, there are exceptions, problems, and obstacles to successful implementation that should be recognized. The first such concern is how to avoid having certain employees always ending up on the bad jobs. Most companies will put their top performers on the most difficult and risky jobs to minimize their potential exposure or losses. With an incentive plan that rewards employees who work on the profitable jobs, where is the incentive to work on the tough, unrewarding projects?

To address this issue, the company must be willing to adjust its profit and schedule categories to give these tough projects a possible positive outcome with respect to the incentive program. For example, if a project has been accepted at cost, then the profit scale could be adjusted to reward employees with a few points if the project does break even. The same could be done for a project with an extremely aggressive schedule; employees could be rewarded if the project was only a few weeks late rather than a few months late. To be fair, the program must be flexible.

A company might also encounter problems with employees who attempt to "work the system" or find ways to artificially increase their incentive without actually completing the required tasks. One of the most effective ways to combat this problem is to keep the program simple. A simple program will have fewer loopholes and hidden formulas that employees can exploit. Also, a program must be flexible to be fair. Thus, if an employee is discovered to have tried to secure an inflated award, then the program must be modified to close that loophole.

A company might also customize its incentive program by rewarding employees who perform well in the incentive plan with public recognition rather than money or trips. For instance, in the design example, each plan cycle Sparks, Inc. could recognize the top three points leaders —the "Sparks Top 3"— with a notice in the company newsletter, dinner certificates, and an engraved plaque to take home. This is a simple way to further recognize outstanding effort as well as create a sense of friendly competition.

Another way to tailor an incentive program is to change the payout frequency. Rather than yearly, the program could be completed biannually or even quarterly. Granted, this will increase the costs of administration to some degree, but employees will receive positive reinforcement more frequently. The frequency of payments will likely be determined by company size and resources rather than a wish to link pay to performance.

As a rule, payouts should not coincide with a Christmas or end-of-year bonus. Employees typically see these types of payouts as "gifts" from the company and not as incentive payouts for outstanding performance. By avoiding confusion between incentive pay and other bonuses, a company will help its employees learn that they receive incentive pay only after productive work performance. The opportunities for specific tailoring can be as varied as necessary to accomplish the goals as set forth by the company. The problems and suggestions described above are relevant to incentive programs in large, medium, or small companies. This is not always the case, especially for many contractors, where companies vary from one- and two-person operations to multinational giants, and no one incentive program design can fulfill all of their needs.

The design example in this chapter was based upon a medium-sized company with revenues of around $6 million a year. Medium-sized contractors share many similarities with both smaller and larger contractors. If there is a common distribution of work, it stands to reason that contractors of varying sizes share common labor needs. Therefore, the question is really one of scale rather than of need, and the incentive program should be adjusted according to the scale of each contractor.

One of the greatest differences in small contractors versus large contractors is how well they know their workforce. A larger contractor that travels from location to location, each time establishing a new dialogue with the local union or workforce, faces different challenges than those of a smaller contractor who works only locally with employees who have been with the company for the last 10 years.

The larger company will not be able to make highly individualistic evaluations because, on an eight-month job with 200 field personnel, time will not allow it. Also, a larger company may not anticipate having work in a certain geographic area very often; thus, the goals of its program are to attract the better workers initially and reward them in such a way that should the company return, the outstanding field labor will want to work for it again. A smaller company will not require the "big bang" similar to the larger company; rather, it will use the program to improve employee relations over time. So the issue of difference in contractor size is more aptly considered a question of the scale of the incentive payment and the complexity of the company's program. To expand upon this statement, recommendations for small, medium, and large companies are outlined in the following sections.

Small to Medium Contractors

Small to medium contractors should increase the number of individual-focused categories to highlight the differences between the fair, good, and outstanding members of the workforce. With fewer employees, greater individual review is possible, and this will keep the program from issuing a similar incentive payment to each employee. Payments of roughly equal value will increase the employee's perception of the payout as a gift and not as a reward for a specific performance. For example, small contractors can more easily implement categories related to customer satisfaction, attendance, or appropriate attire.

Categories such as "years of service" are more useful and rewarding for small and medium contractors. When a company has a more stable workforce, it can use the incentive program to hold on to good employees and attract new ones by providing additional compensation for additional years of service.

When providing incentives to employees with specific skills or certifications, smaller-market contractors are better able to determine each employee's abilities. As a result, smaller contractors can more easily administer incentive programs and can provide skilled field personnel with strong incentives to work for a company with an incentive plan. This will allow for easier administration and stronger incentives for the skilled field personnel to work for a company with an incentive plan.

Medium to Large Contractors

Incentive programs should be simple and focus on project-driven categories, such as profitability and schedule. Contractors too large to know their workforce well must give up individual-focused incentive categories in favor of team-focused categories. The project-driven incentive program will create a pool of money from which employees are awarded shares based upon team performance. This will keep administration to a minimum, will be easily understood by workers, and may supply a large incentive —a "big bang"— that will remind the workforce of the benefits of working for the company.

A larger company is likely to have a formal employee review or evaluation process already in place. This will make the administration of an incentive program less demanding, and the design of the incentive system should work hand in hand with the existing evaluation system. After all, if an evaluation procedure already exists, why invent a different procedure for the incentive plan?

In summary, the incentive plan, if properly matched to a company's goals, should be able to adjust to the difference of scale between smaller and larger contractors. While some amount of tailoring can produce a more effective program, as suggested above, any program is potentially viable with the proper allocation of resources. With attention to simplicity and fairness, modifying an existing program to work for contractors of varying sizes should not require a major overhaul or special clauses.

IMPLEMENTING AND MONITORING THE INCENTIVE PROGRAM

The design has been completed, and the employees have been informed of the evaluation criteria and the incentive plan structure. The next step is to implement the program. We will illustrate this process with an example.

Example 8.1: Alice is a foreman for Sparks, Inc. (our sample company) who has been on profitable projects and has put forth a better-than-average effort in most categories during the last incentive program plan cycle (1 year). She has worked for 2,000 hours during the plan cycle. The data for the profitability and schedule categories are shown in Table 8.6. Notice that, given the weight factors of 5.0 for profitability and 5.0 for schedule, Alice would have gotten 10 points (5.0 + 5.0) if her projects would have performed as expected in terms of profitability and schedule compliance. However, she is actually getting 10.4, as a result of 5.8 points for profitability and 4.6 for schedule. Therefore, she is recognized for working on projects that were more profitable than expected with an additional 0.8 points above the 5.0 baseline. She is also getting 0.4 points deducted from the 5.0 baseline for schedule compliance for working on projects that were a little bit behind schedule.

Table 8.7 shows the safety, job tasks, and marketing categories. Sparks, Inc. recognizes Alice with 2 points for every safety meeting attended, 2 points for every additional safety class, and deducts 2 points for every accident that results in lost time. Alice is also receiving 3 points for every assigned task that was successfully completed (see Table 8.2) and 1 point for every presentation, successful presentation, unpublished project, and successful unpublished project. In summary, Alice receives 18 points in the categories of safety, job tasks, and marketing.

Table 8.6 Profitability and Schedule Data for Alice

Item	Job 1	Job 2	Totals
Profitability:			
Total Gross Profit	65,000	110,000	175,000
Planned Gross Profit	54,000	100,000	154,000
Incentive Pool	11,000	10,000	21,000
Hours on Job	1,000	1,000	2,000
Weight Factor	5.0	5.0	
Points	3.0	2.8	5.8
Schedule:			
Planned Schedule (days)	110	120	230
Actual Schedule (days)	125	125	250
Weight Factor	5.0	5.0	
Points	2.2	2.4	4.6
Profitability and Schedule:			
Total Points	5.2	5.2	10.4

Table 8.7 Safety, Job Task, and Marketing Data for Alice

Item	Totals	Points
Safety:		
Number of Safety Meetings	4	8
Additional Safety Classes	1	2
Number of Lost Time Accidents	1	-2
Points		8
Job Tasks:		
Number of Required Tasks Satisfactorily Completed	3	9
Points		9
Marketing:		
Number of Presentations	1	1
Number of Successful Presentations	0	0
Number of Unpublished Projects	0	0
Number of Successful Unpublished Projects	0	0
Points		1
Safety, Job Tasks, and Marketing:		
Total Points		18

Table 8.8 Discretionary Categories for Alice

Item	Totals	Points
Education and Training:		
Certifications	1	1
Years of Service with Company	6	2
Outside Training	0	0
Points		3
Attire:		
Appropriate Attire Worn	Often	1
Points		1
Client Feedback:		
Feedback	Positive	1
Points		1
Education, Training, Attire, and Feedback:		
Total Points		5

Table 8.8 shows that Alice has completed one certification and that she is getting 1 point in recognition for this education. In addition, she is getting 2 points for her six years of service for the company (see Table 8.3), 1 point for wearing appropriate attire (see Table 8.4), and 1 point for positive client feedback (see Table 8.5) for a total of 5 points in the discretionary categories.

Finally, Table 8.9 shows the calculation of Alice's bonus. The total number of points obtained by all field employees is 105. The total incentive pool is $21,000. However, only 25% of this pool goes to field employees (see Table 8.1), for a total of $5,250. Therefore, by dividing $5,250 by 105 points we obtain the monetary value per point, which is equal to $50. Since Alice has 32.4 points (10.4 + 18 + 5), her share of the filed incentive pool is $1,670. As discussed in the design section, the plan has a target incentive of 5% for high achievers. Alice will receive a 3% bonus for this plan cycle, which is reasonable based on her job performance.

The implementation of an incentive plan should involve free and open communication about the plan to all employee-stakeholders, who should clearly understand what actions will secure the best incentives. This will help the employees to see how accepting responsibility for the company goals will be rewarded.

A company can monitor the plan in several different ways and can use basic statistics to evaluate the program. In the case of an incentive program using points, the program administrator will have point totals for all the employee-stakeholders after the completion of the evaluations. This information can serve a variety of purposes within the company. For instance, assuming the evaluation

Table 8.9 Calculations for Alice's Bonus

Total Points for all Field Employees in the Plan	105
Total Incentive Pool	$21,000
Field Incentive Pool (25%)	$5,250
Value per Point	$50
Alice's Points	33.4
Alice's Bonus	$1,670

criteria do not change, an increase in total points for any category indicates improvement in that field. With respect to profitability, if the employee evaluations produce 100 points one year and 125 points the next year and the company profits are rising, then the company can be confident that this category is correctly measuring profitability.

Similarly, if a company does not usually track data in any of the evaluation categories, the incentive plan will provide a ready-made method to do so. One example is the attire category. If the company never measured compliance with its uniform policy before the incentive program, then it will be able to measure such compliance in future plan periods. In the design example, if the total points awarded for this category rise and the evaluators are consistent in awarding points, then the system has shown that the employees are making a better effort to dress appropriately.

The basic statistical tools of means, quartiles, and standard deviations are helpful in monitoring the information obtained through the evaluations. These functions are available in most spreadsheet programs. Statistical tools provide several checks and balances.

Means:

- Identify the performance of the average employee.
- Verify that the average point award is equal to the planned average point award.
- Verify that each category has the desired weight.

Quartiles:

- Identify which employees are performing in the upper and lower quarters as compared to their peers.
- Offer companies a method of identifying employees in the upper quarter so that they can offer these employees special recognition or

Table 8.10 Sample Historic Data

Profitability Points	Year 1	Year 2	Year 3	Year 4	Year 5	Year 6
Employee 1	3	3	2	3	3	2
Employee 2	2	3	3	3	4	2
Employee 3	4	3	4	3	3	4
Employee 4	4	4	6	4	5	5
Employee 5	2	4	6	2	2	5
Mean	3.0	3.4	4.2	3.0	3.4	3.6
Standard Deviation	1.0	0.5	1.8	0.7	1.1	1.5

Note: Overall mean = 3.433, overall standard deviation = 1.223,
90% confidence interval = 3.054 to 3.812

even an increased payment if employees maintain this level two or more years in a row.

- Verify that the range of scores is consistent with the intent of the design and that there are not unexplained gaps in the point totals.

Standard Deviations:

- Show the variance and spread of each category and the overall point totals. Companies can then determine whether the range is acceptable and can use the standard deviation in year-to-year comparisons to determine whether the employees are performing more or less similarly.

A more advanced statistics tool that can be employed is the confidence interval. To test whether the numbers for a category or the incentive program as a whole are improving or just undergoing normal statistical fluctuations, a company might look to confidence intervals. An example of using confidence intervals to establish a statistical change in a category is presented in Tables 8.10 and 8.11.

As shown in Table 8.10, at the 90% confidence level, the test statistic in the profitability category for the previous six years is between 3.054 and 3.812 points. According to Table 8.11, for Year 7's point totals, the 90% confidence interval is 4.155 to 6.244 points, which does not overlap the previous six years' data; thus, we can conclude with 90% confidence that Year 7 exhibited an increase in profitability points as compared to the previous six years.

Table 8.11 Sample Data for Year 7

Profitability Points	Year 7
Employee 1	5
Employee 2	5
Employee 3	7
Employee 4	5
Employee 5	4
Mean	5.2
Standard Deviation	1.1
90% Confidence Interval	4.155 to 6.244

PRODUCTIVITY AND THE INCENTIVE PROGRAM

After reading through the various categories and evaluation criteria put forth in this chapter, some readers are probably asking: what about productivity? In fact, many companies would list their primary goal in instituting an incentive plan as increasing productivity.

The categories of profitability and schedule will not improve unless productivity goes up. If the estimating department produces accurate estimates, then an increase in profitability and schedule compliance can be attributed to an increase in productivity.

If the company has a track record of successfully measuring and evaluating productivity, then it can use productivity as an evaluation criterion. Contractors who perform repetitive or linear tasks will find adding a productivity category to be more feasible. If a company agrees that increases in profitability and decreases in the schedule are the results of increases in productivity, then the next step is to prove it. Most companies, if not all, monitor productivity, so they can use these productivity measurements to assess the effectiveness of the incentive program.

The previous section discussed basic statistical methods applicable to an incentive program. Those same methods can be used by the contractor to track the correspondence, or lack thereof, between productivity and the introduction of the incentive program.

If the use of statistics is not acceptable, there is a simple and direct method of plotting productivity that will highlight improvements. The contractor should choose one of the productivity measurements and plot the information on a graph. Then he or she should draw control lines that match the peak and minimum productivity outputs during the last few years.

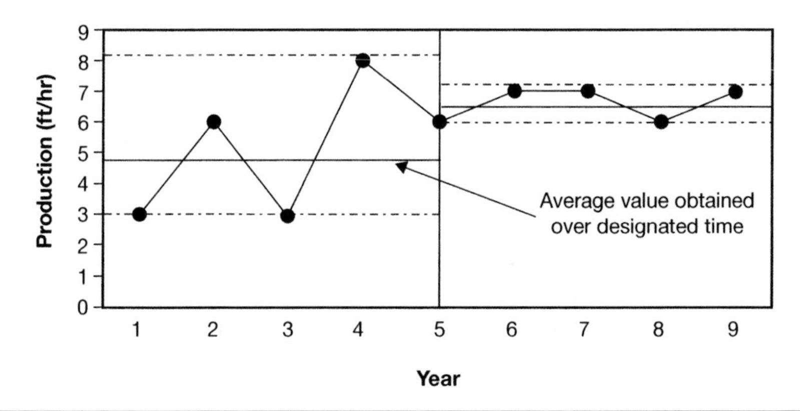

Figure 8.2 Example of decreased variability due to incentive program implementation in year 5.

After the incentive program is instituted, then the contractor can continue to plot this information. If the contractor notices that the control lines have narrowed, this means that the productivity measurements have become more predictable, which implies that there is less variance as shown in Figure 8.2. This can be one positive outcome of the incentive program. Productivity has not necessarily increased, but since the employees know what is expected of them, they work more predictability. This can be an asset to the estimating department, which can now have more confidence in its estimates and can reduce contingencies when bidding jobs.

The other positive outcome, an increase in productivity, can be seen when the lower control line for the productivity numbers, with the incentive plan running, begins to reach the average productivity performance prior to the start of the program as shown in Figure 8.3. The best-case scenario is when the lower control line for the productivity numbers, after the program is implemented, is then above the upper control line prior to the program as shown in Figure 8.4. This would suggest that productivity has improved to such a degree that even normal fluctuations in productivity cannot explain the great increase, which would suggest that the incentive program is a major contributor to the increase.

Those familiar with hypothesis testing can use it to gather statistical proof of the same kind shown in these graphs. The visual representations shown in the figures are meant as a quick check method contractors can use when monitoring the performance of their incentive plans.

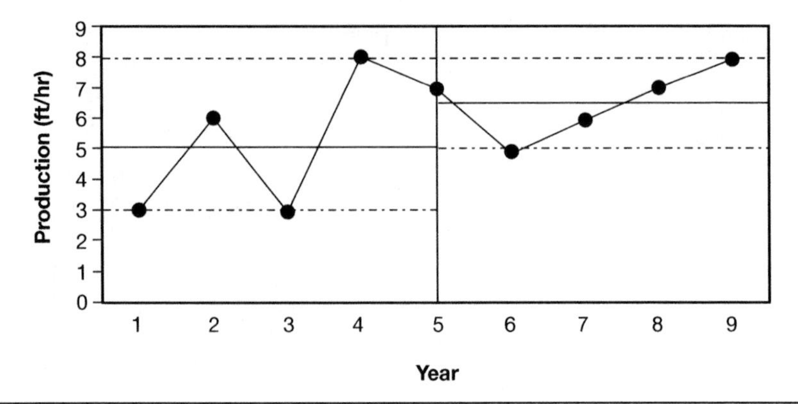

Figure 8.3 Example of noteworthy productivity increase due to incentive program implementation in year 5.

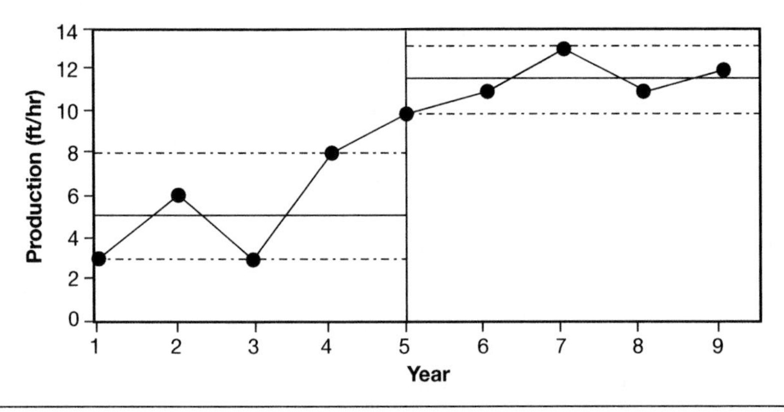

Figure 8.4 Example of significant productivity increase due to incentive program implementation in year 5.

CONCLUSIONS

The incentive system design for Sparks, Inc. is based upon other successful incentive-based programs from the contracting industry and other industries. While the design is unique, it is not unprecedented. To check the functionality of the design presented in this Chapter, we obtained actual operational data from a medium-sized contractor to do a hypothetical run of this incentive system.

The designed incentive system was run for three actual employees from the medium-sized company. In the interest of protecting that company's financial and operational data, the names and locations of the jobs and employees are not provided. Data required for the profitability and schedule, marketing, and discretionary categories for the designed system were not immediately available. In the interest of giving a full view of the system's potential, information was generated in these categories for these three employees to complete their evaluations.

The three employees worked on a variety of projects, where one of the projects was extremely successful, the other two only slightly so. The three projects occurred during the same time period, and each job lasted approximately one year. The three employees where chosen because they worked a substantial number of hours, with each employee logging approximately nine months of labor. The data used for this analysis do not represent average jobs. The sample jobs were chosen from jobs that were successful as a result of high labor productivity. This restriction in the data selection shows the earning potential of high-performance employees.

The results showed that Employee 2 received the most points and the highest bonus ($5,369) as the result of his track record of profit as well as his safety and education record. This would correspond to a bonus of approximately 11% of his base wage (assuming a $50,000 base). This is a healthy incentive payment, which was expected, given that the projects worked on were very successful. The second highest bonus went to Employee 3, with a bonus of $4,679. Finally, Employee 1 received the lowest bonus, at $4,162, because he did not work many hours on the most successful project.

These data highlight the potential payouts that can result from highly successful projects. These three employees received incentive payments ranging from 8 to 11%. An 8% increase to wage earnings is good motivation.

Arguments abound that increased pay does not correspond with increased effort; however, in a study of individual incentive programs, 24 out of 25 companies reported success in achieving their number one goal of increased productivity (Peck 1995). Also, given the choice between one company that rewards employees for accepting company goals and one that simply pays the base wages, the question of employee motivation becomes a question of employee selection. Just as companies seek to offer unique services, they will need to offer unique compensation to their field employees. An incentive plan like the one described in this chapter will ensure that a company can recruit and retain top fieldworkers and thereby can provide its customers an incentive to use its services in the future.

9

JOB STRESS IN CONSTRUCTION SUPERVISORS

Dr. Terry L. Stentz, *University of Nebraska-Lincoln*
Kathleen J. Spanjer, *University of Nebraska-Lincoln*
Megan P. Ewer, *University of Nebraska-Lincoln*
Blake Wentz, *University of Nebraska-Lincoln*

INTRODUCTION

There is ample evidence to show that problems related to stress are the number one chronic health concern in the United States, accounting for an estimated 75% to 90% of all visits to primary care physicians. Moreover, occupational stress is considered to be the leading source of stress for American adults directly related to health problems such as back pain, feeling fatigued, muscular pains, and headaches. The cost of job stress in U.S. industry is estimated at over $300 billion annually as a result of accidents, absenteeism, employee turnover, diminished productivity, direct medical, legal, and insurance costs. Many in the industry would agree that construction is one of the more stressful industries for supervisors and managers. A sizable share of the common work stressors in construction management could be mitigated through stress and coping education and appropriate management reinforcement and support. Therefore, understanding more about the various sources of work and life stress for construction supervisors is an obvious area of interest.

The first objective of this study was to identify the sources of job and life stress in a convenience sample of electrical construction supervisors. The second objective was to evaluate relationships between job and life stressors and personality type that could assist contractors in addressing stress management strategies and skills for better job performance and life adjustment. The third objective was to use the results of the study to make recommendations to contractors related to lowering job and life stressors in support of improved job performance and life adjustment.

A survey battery consisting of five validated and widely accepted psychological instruments and one demographic and job information questionnaire were administered to the volunteer subjects whose job description was electrical foreman or electrical superintendent. The raw data for each instrument were analyzed and transformed into a series of scales according to the requirements and criteria of each psychological instrument. The data analysis included descriptive statistical analysis, means comparisons, and correlation analysis. The five psychological instruments used in this study were as follows:

- Job Content Questionnaire (JCQ)
- Occupational Stress Inventory—Revised Edition (OSI-R)
- Life Stressors and Social Resources Inventory—Adult Form (LISRES-A)
- Myers-Briggs Personality Type Indicator (MBTI)
- Personal Behavior Pattern Questionnaire

The demographic and job information questionnaire was designed by the research team to gather data related to age, gender, marital status, children, job title, job experience, type of construction project or work, span of control, commuting distance, and technical and nontechnical education levels.

DEMOGRAPHIC QUESTIONNAIRE

Survey packages were mailed to a convenience sample of 83 electrical construction supervisors of which 75 responded (90% response rate). Each respondent's current job title was either electrical construction foreman or electrical construction superintendent. The response subject sample represented permanent residency in 62 cities from 17 states. Three out of the 75 subjects (4.0 %) had temporary residencies due to construction project location. Three of 75 respondents were female (4%) and 72 were male (96%).

The mean (average) subject age was 45 years old with an average of 23 years of work experience in the electrical industry, and 9.5 years with their current employer and place of employment. The highest level of formal education for approximately 33% of the subjects was completion of grade 12 (high school graduate), 48% had completed some college but no degree, and 19% had less than a high school diploma. Approximately 73% of the subjects had completed the IBEW Joint Apprenticeship Training Program.

Approximately 80% of the subjects were currently married, 4% single, 11% divorced or separated, and 5% no response. The average number of children living with a subject was 1.35 with approximately 62% of the sample having at least one child and the remaining sample having two or more children living with them. The average age of the children living with subjects was close to 12 years of age.

The average number of hours worked per week was 42.5 hours. All subjects indicated that they work at least 40 hours/week, and some reported a maximum of 70 hours/week. Approximately 88% of the sample worked a five-day workweek and 12% more than a five-day workweek; 96% worked first shift and 4% worked other than first shift. First shift was defined as a workday starting between 7 a.m. and 8:00 a.m. and ending between 4 p.m. and 5 p.m.

The subject sample represented current work on 94 construction projects in 46 cities from 15 states. The average span of control for each subject was 17 subordinates with several subjects reporting a maximum of 125 to 135 subordinates. A broad range of construction projects and types were represented, with commercial/industrial construction predominating.

JOB CONTENT QUESTIONNAIRE

The Job Content Questionnaire (JCQ) is a job assessment instrument designed to measure psychological demands, decision latitude, social support, physical demands, and job insecurity, as shown in Table 9.1. Each job content criteria requires a series of scaled responses indicating the degree of presence or nonpresence of the job content criteria features or characteristics shown in Table 9.1. In accordance with the demand/control model hypothesis (Karasek and Theorell 1990), job strain is predicted when the psychological demand is high and worker's decision latitude is low with lower levels of social support, which increases stress levels and resultant strain. Conversely, high levels of job motivation can exist even under high psychological demands if high levels of decision latitude are present. Low job demands with low decision latitude can lead to negative learning or gradual loss of previous acquired skills.

The vast majority of the subjects stated that they do not suffer from stress over job security, and most are satisfied with their job. Almost 70% of the subjects stated that they would take the job again if they had a choice, and over 85% of the subjects had no intention of looking for a new job in the next year.

Table 9.1 Job Content Questionnaire Criteria Description

Criteria	Descriptors
Psychological Demand	General Psychological Demands
	Role Ambiguity
Decision Latitude	Skill Discretion
	Decision Authority
	Underutilization
Social Support	Coworker Support
	Supervisor Support
Physical Demands	Physical Exertion
	Hazardous Conditions
Job Insecurity	General Job Insecurity
	Global Job Insecurity

Over 90% of the subjects stated that they must be creative to complete their jobs. Over 90% of the subjects also stated that they are allowed freedom to complete their tasks, and they are given decision-making authority by their employer. These subjects also stated that they have a great say in what happens in their job, they are allowed to do different things, and they have the opportunity to develop their own special ability. These subjects also stated that their ideas had a significant influence over group decisions and may be considered for company policy.

The subjects reported that they are exposed to some hazardous conditions in their job, and that their job requires quite a bit of physical exertion. Over one third of the subjects reported some physical pains such as back and neck pain related to their job. One third of the subjects also reported having difficulty with their sleep patterns, such as having trouble falling asleep or staying asleep, or even waking up too early in the morning. The subjects also reported being easily irritated over unimportant items and other stress reactions.

OCCUPATIONAL STRESS INVENTORY

Occupational Stress Inventory—Revised Edition (OSI-R) is an instrument to measure three dimensions of occupational adjustment: occupational stress, psychological strain, and coping resources. Each of these dimensions is measured by a set of scales of attributes, shown in Table 9.2. Each of the psychological constructs (stress, strain, and coping) is measured with scaled responses for each of

Table 9.2 Occupational Stress Inventory (OSR-I) Description

Criteria	Descriptors
Occupational Stressor	Role Overloaded
	Role Insufficient
	Role Ambiguity
	Role Boundary
	Responsibility
	Physical Environment
Psychological Strain	Vocational Strain
	Psychological Strain
	Interpersonal Strain
	Physical Strain
Coping Resources	Recreation
	Self-Care
	Social Support
	Rational/Cognitive
	Coping

the construct components shown in Table 9.2. Stressors are sources of stress to which a person reacts in the form of strains, with strains mitigated by different sources and types of coping mechanisms, behaviors, or activities.

The most serious occupational stressor recorded by the subject sample was role responsibility. Over 64% of the subjects recorded role responsibility as the primary stressor. The majority of the sample indicated that their role boundary and the physical environment were average with expectations for their type of work. Role insufficiency was not of significant concern, but concern for role overload was rated as the second highest stressor.

A majority of the subject sample stated that the strains identified in the survey were average or below average level. Interpersonal strain was the highest strain identified, with only 12% of the subjects stating that their jobs had a mildly high level of interpersonal strain. The next highest strain was physical strain identified as significant by only 9% of the subject sample.

The use of coping strategies surveyed was rated as average by a majority of the subject sample. The level of social support was identified as the most frequently used and effective method for coping (31%) and recreation second (26%). The average coping deficiency was a 10% deficit for the subject sample with self-care level being the highest, most frequent, and effective for 16% of the subject sample.

Table 9.3 Life Stressors and Social Resources Inventory Criteria Description

Criteria	Descriptors
Life Stressors	Physical Health
	Home and Neighborhood
	Financial
	Work
	Spouse or Partner
	Children
	Extended Family
	Negative Life Events
Social Resources	Financial
	Work
	Spouse or Partner
	Children
	Positive Life Events

LIFE STRESSORS AND SOCIAL RESOURCES INVENTORY

Life Stressors and Social Resources Inventory-Adult Form (LISRES-A) is used to measure two criteria: stable life stressors and social resources. This questionnaire includes nine sources of life stressors and seven sources of social resources. These measures are listed in Table 9.3.

The most important source of life stress for the subject sample was children: 34% of the sample stated that their children produced an above average level of life stress, and 17% indicated an average level of stress. Work was identified as the next highest life stressor by 31% of the sample, followed by spouse stress (23%). The lowest ranked source of life stress was identified as financial, which was rated as average for only 7% of the sample.

In the area of social resources (available for coping with life stressors), 67% of the subject sample indicated that their financial level was an above-average coping resource. The next highest ranked social resource was positive life events for 29% of the sample. The most important deficiency of social resources support was identified as children, with 58% of the sample rating this resource as below average. The extended family was ranked second with 52% of the sample, followed closely by spouse level of support with 42% of the sample rating this social resource as below average.

Table 9.4 MBTI Perception/Information Taking and Decision-Making Dimensions

Perceiving/Information Taking		Decision Making	
Sensing (S)	**Intuition (N)**	**Thinking (T)**	**Feeling (F)**
Perception favors clear, tangible data and information	Perception favors abstract, conceptual, big picture and represents imaginative possibilities	Makes decisions in an objective, logical, and analytical manner	Makes decisions in a global, visceral, harmony- and value-oriented way; pays attention to the impact on other people

MYERS-BRIGGS PERSONALITY TYPE INDICATOR

The Myers-Briggs Type Indicator (MBTI) is an instrument used to self-rate and classify personality preference types. The model holds that people differ from one another in four different ways or dimensions. The first pair of personality dimensions is related to mental preferences as shown in Table 9.4. This pair of dimensions reveals how people perceive or take in information. Those who favor Sensing (S) perception pay more attention to clear, tangible data and information that fits in well with their direct here-and-now experience. By contrast, those who favor "Intuition" (N) perception are drawn to information that is more abstract, conceptual, and big-picture, and that represents imaginative possibilities for the future.

The second pair of dimensions, also shown in Table 9.4, identifies how people form judgments or make decisions. Those who prefer Thinking (T) judgment have a natural preference for making decisions in an objective, logical, and analytical manner with emphasis on tasks and results to be accomplished. Those whose preference is for Feeling (F) judgment make their decisions in a somewhat global, visceral, harmony- and value-oriented way, paying particular attention to the impact of decisions and actions on other people.

There are two other mental preferences that are part of the Myers-Briggs model as shown in Table 9.5. One of these preferences pertains to energy consciousness Extraversion (E) versus Introversion (I). The other one is life management orientation Judging (J) versus Perceiving (P). Thus the permutations of these four preferences result in 16 personality types that form the basis of Myers-Briggs model and the MBTI instrument.

Of the 16 personality types, the largest proportion (31%) of the sample turned out to be the ISTJ (Introversion/Sensing/Thinking/Judging) type personality.

Table 9.5 MBTI Energy Consciousness and Life Management Personality Dimensions

Energy Consciousness		Life Management	
Extraversion (E)	**Introversion (I)**	**Judging (J)**	**Perceiving (P)**
Outgoing	Quiet	Tries to control life environment	Flexible, adaptable with environment

These individuals tend to be quiet, logical, and objective. They rely on tangible information and data to try to control their lives and their jobs and are most comfortable with closure in terms of decision-making. Doing one's "duty" is very important to this personality type.

The next largest group (17%) was the ENTJ (Extraversion/Intuition/Thinking/Judging) type personality. These individuals are different from the previous group in that they are outgoing, relying primarily on conceptual data or information and taking the big picture more into account when making decisions. Similar to the ISTJ personality type they are logical and analytical in attempting to control the situation. Being a "natural leader" best characterizes this personality type.

The only other personality type represented by the subject sample was the ESTJ (Extraversion/Sensing/Thinking/Judging) personality type. These individuals are just like the ISTJ with the major exception that they are much more outgoing but still requiring clear data, thinking and acting logically and desiring to control the situation. Being a good "administrator" best describes this personality type.

PERSONAL BEHAVIOR PATTERN

Personal Behavior Pattern (Type A–B) is a simple self-measure of personal behavior type pioneered in the prospective health and mortality study of Harvard medical students after graduation. The results of this research revealed two distinct types of individuals: one with high levels of internal stress and the other with much lower levels of internal stress. It was found that Type A personalities had higher rates of stress-related health problems and mortality after age 40 than Type B. The description and differences of personality between Type A and Type B are shown in Table 9.6.

More than 84% of the subject sample was classified as Type A. As noted in Table 9.6, Type A individuals are driven, aggressive, competitive, constantly under internal stress and time pressure, impatient, and often unable to cope with leisure time. These individuals are very detail oriented, concerned with time deadlines, and quite comfortable with quantified values to evaluate their performance and the performance of others.

Table 9.6 Personal Behavior Pattern Classification Description

Personality Type	Typical Characteristics
Type A	Is always moving
	Eats rapidly
	Does two things at once
	Cannot cope with leisure time
	Is obsessed with numbers
	Measures success by quantity
	Is aggressive
	Is competitive
	Constantly feels under time pressure
Type B	Is not concerned with time
	Is patient
	Plays for fun, not to win
	Relaxes without guilt
	Has no pressing deadlines
	Is mild-mannered
	Is never in a hurry

CONCLUSIONS

Based on the results of this study, the following general recommendations are provided for contractors to consider in designing and executing a plan to help their supervisors better understand and cope with job and life stressors for improved productivity and longevity:

- Evaluate, discuss, and redesign supervisor jobs to "right-size" the span of control. Each supervisor should be directly responsible for no more people than he or she is comfortable and effective with while keeping the chain of command as "flat" as possible (fewest levels of authority needed to do the job and maximize communication). Direct input from experienced supervisors is required to effect this change.
- Provide stress and coping education and skills development training for supervisors and encourage (reward) participation and stress-lowering behaviors. Contractors could benefit from taking this training and modifying their own behaviors right along with their first-line construction supervisors. Joint participation in this training and behavior modification experience could add additional credibility and efficacy to these changes as well as foster better communication between upper management and supervision. Qualified professionals can and should be utilized for establishing workplace stress and coping programs.

- Provide supervisors with sleep, sleep hygiene, and sleep disorders information. Emphasize the resources available in the community to get specialized evaluation and treatment of chronic sleep disorders and chronic insomnia. Emphasis should focus on the value of good sleep to increased productivity, better decision making, and overall health and stress hardiness for supervisors. Qualified professionals can and should be used for employee training and especially for referral, evaluation, and treatment of sleep problems.
- Provide information on community resources available to supervisors for parenting education and child-rearing skills development. In reality, businesses hire families. Strong families foster strong, well-adjusted, productive employees. The strengths from home come to work and vice versa. Since the major stressor revealed by this study was related to children and life partners (spouses), supervisors and others in the organization could benefit from family support activities sponsored by the company, community marriage counseling resources, spouse support groups, and other family support and child development resources available outside of work. Contractors who recognize and acknowledge these types of adjustment and developmental problems and how they relate to supervisor stress reduction and increased productivity are leading by example in connecting outside support with the important people who make the work happen every day on the jobsite. Professional family support and parenting educators can and should be used for this type of intervention.
- It is recommended that Type A-Type B behavior and MBTI personality preference type training (or similar) be used in the management development program for supervisors and managers. Increased awareness, knowledge level, and self-recognition of various characteristics, behaviors, personal preferences, communication styles, and consequences of these various behaviors related to individual and organizational stressors is more than half the battle in reducing stress and increasing coping skills and understanding in individuals as well as the organization. Qualified professionals can and should be utilized for these assessments and for individual or group interventions warranted by these assessments.

10

RECOMMENDED PRACTICES FOR PRODUCTIVITY ENHANCEMENT

Dr. Jerald L. Rounds, *University of New Mexico*

INTRODUCTION

The purpose of this chapter is to present the results of a study about productivity enhancement focusing on labor efficiency. The goal of the study was to stimulate performance improvement within the construction industry through enhanced productivity. The project was based upon the premise that the first step in improving performance is to discover and disseminate existing means for productivity enhancement so that proven approaches used on a limited basis can be broadly implemented throughout the industry. To accomplish this goal, three objectives were developed: (1) learn from industry leaders what they feel are the keys to their success so that these keys can be disseminated broadly across the industry, (2) collect data on labor practices within the industry as a means to better understand both problems and solutions in improving productivity, and (3) determine areas in which contractors' field efforts can be profitably invested to study industry problems and practices as a means to improve labor productivity.

Eleven focus group workshops were convened with leaders of the construction industry to find out both what they feel their companies are doing right and what areas of opportunities they saw for future studies. Participants represented each of the 10 districts of NECA and a cross section of labor unions. Two of the 11 workshops were with National Joint Apprenticeship Training Program trainers. These groups were unique in the sense that participants were not asked to come prepared with concepts, since they do not represent companies. Their contribution was to provide the viewpoint of labor concerning what affects labor productivity. Their comments were well aligned with those of other workshop participants, confirming that both labor and management know a good deal about what affects productivity and that they are generally in agreement. Two of the contractor focus groups were carried out with large groups of 50 to 100 participants who were divided into smaller groups of 8 to 10 people.

FOCUS GROUPS

Focus groups were moderated by a facilitator who ran the program as a workshop with each participant expected to enthusiastically contribute. Each meeting lasted up to one day and adhered to the following agenda:

- *Program and participant introductions.* Introductions were particularly important to set the stage for strong discussions. They built a feeling of confidence and a level of understanding of the background from which ideas came.
- *Concept sharing.* Each participant shared a concept that had been proven effective within their company. Brief discussions took place to ensure all other participants had a clear understanding of the concept. This cycle was repeated several times to allow participants to place additional concepts on the table.
- *Team concept analysis.* The concepts were arranged into groups with similar characteristics (i.e., crew application, project application, administrative application). The focus groups were split into teams to study the concepts in each group, asking such questions as: (1) How can this concept be improved? (2) How valuable is this concept? (3) Where is this concept more applicable? (4) Where is this concept less applicable?
- *Team concept feedback.* Each team summarized their analysis for the group at large.
- *Program wrap-up.* The program closed with a discussion of how each participant could return to their office with at least one new concept

to implement. Information gathered from the workshop was summarized by the facilitator and a summary of the most valuable concepts was mailed to each participant.

Focus group participation was limited to not more than 12 participants in order to support the extensive discussion essential to successful exchange of ideas. Each participant was asked to bring with him or her at least one productivity enhancement concept to introduce to the focus group. Ideas focused on enhancing labor efficiency as a means to improve productivity, rather than looking at improvements on materials or equipment. Support documentation such as sample forms, report pamphlets, or training materials were provided to the group.

Participants were experienced industry practitioners rather than younger persons with limited experience. Participants came with an attitude of listening and sharing in a positive environment. Since the concepts were collected to create an Idea Cache that contained ideas from a number of focus groups, additional information was requested from some contributors by the workshop facilitator sometime after the workshop to enable complete analysis and thorough reporting on the concept for the Cache.

The primary product of the focus group workshops was a set of productivity concepts provided by participants that they felt represented keys to their operations. Because some contractors felt that these key concepts were proprietary and not to be shared with competitors, the idea of sharing was marketed on the basis that their key concept would be "invested" for a return of general commentary from other workshop participants on how it could be improved, as well as the gaining of many other productivity key concepts from other workshop participants.

Many of the concepts deal with similar topics because contractors tend to strongly agree on those areas that are most important to their operations. Therefore, concepts were group in four categories: (1) tracking labor units, (2) project startup, (3) employee relations, and (4) company management. These concepts are described in the following sections.

PRODUCTIVITY CONCEPTS

Tracking Labor Units

Concept 1: Systematic Estimating Process for Small Contractors

Cost estimating is critical to the success of a construction company. Estimating approaches have traditionally incorporated unique characteristics for each company, though standardization of estimating software tends to play an important part in establishing a more common basis for construction estimating for larger companies.

On the other hand, because investing in a software package requires more of a commitment of both money and time than small contractors may be able to afford, smaller companies, which form the backbone of the industry, still tend to develop their own estimating approaches and are not being led as readily toward standardization.

This concept is focused on the small company where the owner and possibly one or two others take care of the estimating. A systematic approach is important, but it becomes critical when the primary estimator in the company is preparing to step down and needs to train a successor.

The systematic approach to estimating job costs is based upon appropriate forms that lead the estimator through the process combined with a thorough site visit that enables the estimator to tailor the estimate to the specific job. The site visit is particularly critical with the current trend in design to leave out a great deal of detail.

On the site visit, quantity information is gathered and entered into forms. These forms have line items for major work activities. A special template is used in the office to enable consolidation of information. From a consolidation sheet, quantities are moved to a pricing sheet. Material is then priced and extended. Labor units are applied and extended. Finally, summary pricing information is moved to a recapitulation sheet, which summarizes estimated costs for the entire job, and includes general-conditions items and markups, resulting in total job costs. Nowadays, small companies may be able to afford the automation of this process by using PDAs loaded with standard spreadsheet software.

To ensure a successful implementation of this concept, you should have someone overlooking the project when it is done who knows more about the project than the foreman in the field. This is the estimator that carried out the initial site visit, project planning, and estimating. Foremen are provided the quantity detail sheets, but work-hour and pricing information is not given to the foreman. The estimator/supervisor reviews work progress and investment on a weekly basis with the foreman.

Concept 2: Tracking Field Labor Units in a Niche Market

Tracking field labor units is a key to both managing jobs and planning strategically for the company. Typically, much data is gathered, but it is inadequately analyzed and is not utilized to help chart the company direction. This concept proposes focusing operations in well-defined niche areas and limiting data collection to a small number of activities that are key to this specific type of work.

Moving into a new specialty area provides both the need and the opportunity to build a new database containing field productivity information. The data can be used at the job level to control productivity and at the company level to help refine the market by determining the types of work at which the company excels. Key to this concept is visualizing company operations as a collection of specialty areas.

This concept involves gathering accurate field data on a limited scope of work. Too much detail overwhelms users and frustrates field supervisors who are asked to collect the data. The response of supervisors is to generate inaccurate and meaningless data. One solution to the problem is to narrow the data collection to a small number of key activities. The limited amount of data is not overwhelming, and the job of coding work hours to a limited number of activities simplifies the supervisor's job. To keep the number of activities low, each project is considered to fit into a specialty area. Different sets of forms and a different database are maintained for each specialty area. A more complex job may be composed of several specialty areas.

The information management system is built around a series of forms. The weekly timesheet form is set up as a matrix with weekdays appearing vertically and work categories horizontally. On a given sheet, data can be kept for up to five employees for one week. Up to a dozen work category columns correspond to the major activities in the specialty area. All labor must be coded to one of these categories.

Two types of analysis are carried out on the information. The first provides a job production labor analysis. It is based upon information from both the timesheet and the estimate, and shows hours invested for the month, total hours invested to date, and percent of hours invested to date for each work category. It also gives the percent each job category represents of the total job. The supervisor uses this information to monitor work progress as he or she walks the job. The report enables the supervisor to focus on work categories that are a significant portion of the total project and categories that may be behind in production units, either this month or to date.

The second analysis is a labor summary report that calculates the hourly rate, to date, for each work category. Both the estimated and the actual rates are calculated, which enables comparison for evaluating production on a specific job, as well as historic production data that can be used for estimating future jobs. The variable hourly rates reflect the way in which the work is crewed. The rate will vary in the estimate if the estimator identifies a difficult task and plans on assigning more expensive labor to the task. The field rates reflect how the supervisor did, in fact, crew the various work items.

One additional form provides a description of each work item on the timesheet so that field supervisors will know just what work goes into each category. This is a particularly important component of this concept that breaks work down into a limited number of categories, each of which must account for a variety of related types of activities.

For this concept to work, a company must feel comfortable with limited detail on field operations. This is accomplished by dividing all work into specialty areas and by combining similar types of work into a single category. Also important is support to help the supervisor code labor hours to the proper work item. This help could be in the form of training the supervisor in the extreme importance of accurate reporting and is very dependent on not assigning blame when work hours exceed the estimate.

Concept 3: Job Costing Report

Controlling costs at the job is critical to the success of any construction company. All other functions in the company support or complement the field operations, which represent the primary function of the company.

A company may want to consider job costing as the hub of its computerized management system. All relevant information from other parts of the system, such as estimating and accounting, flows into the job costing system. This information can then be used both to manage the job and to provide essential information for the estimators to price out jobs.

The job costing report breaks the job down into a small, manageable number of work activities. A two-part breakdown, labor and materials, is provided for each category. Labor information is provided in both hours and dollars. For both hours and dollars, the analysis provides information on percent used, budget, to date, remaining, and over/under. The average unit labor cost to date ($/hr) is also provided.

As with any job costing system, collection of field data is critical. It is imperative to train field supervisors to fill out time cards accurately, recording the correct number of hours under the correct work item code.

Concept 4: Field Labor Unit Tracking System

To improve productivity, it is necessary to be able to monitor the progress of work under way. The level of detail and the timing of reports varies depending on specific company and job objectives, but adequate detail and close enough timing is required to be able to detect problems at a sufficiently early stage so that effective evasive action can be taken. In addition to time and patience, developing such a system requires an understanding of the type of information required and how that information is to be used.

Development of the reporting system explained here began with five major categories. This provided some information, but not enough. The system underwent expansion until its current state where it contains about 1,000 items. Only those appropriate to a specific job are used for that job. Other items can be added if required, but with 1,000 items, addition of specialized items is not often required.

The system is fully coordinated with the company's estimating system. Selection of appropriate line items for a given job is done during the estimating process. Additional line items can be added by the field supervisor when field conditions vary from those estimated. For instance, if a material is replaced by another, then a new item must be added.

The system tracks both labor dollars and hours, providing five items of information for each: estimated, actual, percent actual/estimated, variance, and current period quantities. In addition to labor units, the system also tracks materials by coding material line items in the thousands to correspond to a specific item. Thus, for labor item 530, the material is tracked in item 1530. In addition, an appropriate payroll system must be in place to enable collection of work hours on a line item basis corresponding to the labor tracking system.

Monitoring the system takes time. The developer of this system began the work when he was a regional manager. He initially monitored the reports for all jobs himself once a week. As the volume of work increased, especially with his elevation to CEO, this became too time-consuming, so a special person was brought in with the single responsibility of monitoring job labor productivity. This physically disadvantaged person had special skills and a background as a journeyman and manager.

A critical part of the system is to ensure that the field data is correct. The person monitoring the system must review field data for consistency and reasonableness. For instance, if excessive work shows up in the "miscellaneous" category, something is wrong. Likewise, if work is coded to work items that cannot yet be active, field data is being erroneously coded. One way to identify improperly coded work items is to check whether material has been charged against the corresponding labor item.

As an example, this system was used early in the development stage to monitor a large job consisting of pulling 7 miles of 500 MCM high-voltage cable. Because the company had little real-world knowledge about how much it cost to actually pull wire but had a "gut feeling" that wire pulling was always bid high, the job was bid using a multiplier of 0.7 on labor units. The system allowed tracking of actual costs, showing that the actual labor units expended were only 52% of the bid labor units. As a result, the system has provided valuable information for bidding future jobs. An additional benefit was that a major claim was resolved in the contractor's favor based upon the detailed data collected.

After this project demonstrated that estimated work hours were at such a variance from actual, various other jobs were reviewed. This review process demonstrated that in jobs where large wire was being pulled, the estimated units invariably exceeded actual units, whereas where small wires were pulled, actual units exceeded estimated. As a result, a further breakdown of the system was required to provide for different sizes and types of wire. In like manner, other major categories like conduit and fixtures also had to be further broken down. As previously mentioned, the resulting system now has on the order of 1,000 line items.

The next step was an expansion to enable coding of nonproductive time. For instance, a line item was added for meetings so that when labor is in safety meetings, time can be so coded, thus improving the accuracy of tracking productive work and separating out nonproductive work hours.

Concept 5: Field Labor Tracking System

A major factor in remaining competitive in construction is management of field activities. For many specialty contractors, field labor could account for as much as 70% of job costs. Field labor is the biggest gamble in construction. Tracking materials is another big problem. This concept focuses on a tool to monitor, and hence control, field labor as well as materials in the field.

Particularly on large projects, some means must exist to get accurate and quick information in a concise format back to field supervisors on the status of both labor productivity and material management. Timeliness is important, since with ongoing projects, delay in getting information will allow the window of opportunity for making adjustments to close. Flexibility is also important because each job is different and both the number and the selection of items to monitor will differ from job to job.

Virtually all companies have systems in main or regional offices that track materials and labor costs, along with the other jobsite costs, and feed these back out to the jobsites. The described concept is complementary to that system.

A tracking system has been developed in-house to be used in laptop computers at the jobsite to monitor both materials and labor productivity. It was designed with a great deal of flexibility so that the level of detail can be adjusted to meet the objectives of a specific job. It has the capability to handle a vast number of work codes. Most of these are built-in; however, others can be added for a specific job. Not all work items will be used in each job. Only those that are needed to meet the specific objectives of the job will be used. Objectives might include looking at specific physical areas on the jobsite or focusing on specific crews. Of course, care must be taken in tailoring the reports for a given job because the more detail that is incorporated, the more time it takes to maintain and use the system.

Field data is entered at the jobsite. The general foreman is responsible for collecting data from all the foremen for entry into the computer on a weekly basis. Labor time is entered off the time cards. Time card data also goes to the main office, but the field system allows collection of labor data and immediate analysis so that delays do not occur waiting for information to come back from the main office accounting systems. Since analysis is immediate, the most data is behind is one week. Field supervisors are provided the traditional accounting data when it comes out, which provides some different information that may prove valuable. The material tracking capabilities enable monitoring such things as the supplier for a specific item, shipping status, and payment status.

Development of a sophisticated program requires significant time and investment. The system detailed here would require a sufficiently large company that could shoulder this initial investment. On the other hand, computer costs continue to go down and high-performance computers can be bought at very reasonable prices even in the form of PDAs. Furthermore, computing literacy is increasing every day. As a result, an employee with a good spreadsheet and a relatively inexpensive computer or PDA could develop a simple labor and/or material tracking program relatively quickly. Once up and running, it can always be expanded as time and demands require. In this way, even smaller companies can benefit from a site-based labor and material tracking system.

Another requirement for implementing such a system is communication to field supervisors. First, they must be convinced that this system is not a threat to them—in other words, that it is not used simply to watch them more closely. Second, they must be shown the value of this system. Finally, they must be taught how to use the system effectively.

Various types of problems can be detected with the system. There are times in a project when labor slows down. They may not have the right directions or tools, or they may be contemplating the end of the job and want to hang on to this job as long as possible. Constant monitoring of the current period labor productivity will show when there is a change in productivity rates.

Another problem may be having the wrong crew doing the wrong job. If a cable pulling crew is not making the estimates rates, it may mean that they are inexperienced. It might be appropriate to switch them to installing conduit and get a more experienced pulling crew to pull wire.

The system is also valuable as a means to collect data from the field for feedback into the estimating system.

Project Startup

Concept 6: Mobilization Questionnaire

A common problem in construction is getting the job started correctly. The project manager/estimator–foreman interface is very important. Much time is lost in the field when field labor and supervision must try to determine details about the job from the prints. The project manager/estimator becomes thoroughly familiar with the job through the estimating process and is responsible for communicating these details to the field supervisor. In this way, field decision making is minimized, which minimizes disruption to the job and potential problems arising out of field decisions that do not align with the big picture of the job. It is the project manager/estimator's job to see that the fieldwork goes the way it was planned.

Fullest advantage must be taken of the time from award to bid to job start in order to organize all information, materials, and equipment so that on day 1, work can start. This includes gathering forms, approvals, the latest set of prints, all changes or modifications, and a copy of the contract with any changes. Field labor wants to do a good job and it is management's responsibility to see that they have everything needed to do so.

Foremen need full disclosure of all aspects of the job, including dollars and/or hours allocated to each work item. One contractor goes over each new job estimate line by line with the foreman before that foreman goes out to the job.

To expedite job mobilization, one contractor has developed a questionnaire of about five legal-size pages that leads the foreman or project manager through the entire set of specifications in order to extract critical details about the job.

Another contractor has developed a project startup checklist that divides the startup procedures into two categories: (1) administrative and (2) tools, equipment, and materials. It then lists each item that must be accomplished to get the job started, and leaves room for short answers to indicate status. Some items, for instance, "Vehicles," simply have blanks to fill in. Under "Rental Equipment," a blank is left to describe the equipment to be rented and a place is left to specify delivery date. Some items simply have a check-off, such as the list of tools to be supplied. Others, like construction drawings, have both a target date and a check-off to indicate when the action called for has been completed.

Well-defined startup procedures and the associated checklists and questionnaires can be used effectively to start any job efficiently. An extensive set of procedures and forms will take some time to develop, but this type of concept can be implemented slowly, stage by stage. Simple procedures and forms can be developed and used on one or two jobs. What works can be used on other jobs, and new procedures or forms can be developed and added to the standard set as needed. This type of concept can always be improved upon.

Concept 7: Foreman Training for New Hire Orientation

One of the characteristics of the construction industry is continual turnover of field employees. Field productivity is strengthened if new employees get off to a good start. Proper orientation is a valuable tool in enabling new employees to start their new jobs productively and safely.

Certain information, such as safety requirements and procedures on the job, must be communicated to each new employee. Other information must be collected from the new employee for bookkeeping purposes. Providing other information to the new employee, such as layout of the site and policies and procedures for running the job, although not required, is important to help the employee rapidly come up to speed on the job.

This concept consists of the implementation of a training program that allows the foreman to better orient the new hire. Some orientation is required in order to gather essential accounting information and to impart important safety information. Since such orientation is a requirement, expansion to a much more comprehensive orientation program will require little investment, but will show significant returns in improving productivity on the current job, and as long as the employee works for the company.

The foreman training program incorporates a number of components:

- It instructs the foreman in the importance and objectives of the new hire orientation.
- It summarizes the information to be collected from new employees.
- It summarizes information to be provided to the new employee when he or she first comes to the job.
- It enables orienting the foreman as to important responsibilities he or she has on the job, as well as ensuring they have the information, certifications, and competencies needed to competently perform the job.

To implement this type of foreman training, the company must first have developed a well-defined new employee orientation program. The program should incorporate attractively developed data on the company, including background, philosophy, policies, and procedures. It should also include forms required to be filled out by the employee to incorporate his or her profile in the company accounting system. Next, the foreman training program needs to be prepared. It should then be presented to all current foremen. They should be invited to comment on how the program can be modified and improved before it is presented to new foremen. Finally, provision should be made to orient each new foreman when they are raised to this new level.

Concept 8: Interdisciplinary Foreman Training

Foreman training is recognized as a critical element in improving labor productivity. The foreman is typically a journeyman who has demonstrated traits the contractor believes will make him or her a good supervisor and was therefore promoted. Regrettably, being a foreman does not require the same skills as being a craftsperson and too often foremen are not trained in supervision. This concept presents a unique, interdisciplinary foreman training program.

In one large metropolitan area, an interdisciplinary foreman training program was developed through the efforts of a number of different labor unions. The basic course consists of six 3-hour modules focusing on content of interest to construction supervisors. Each module is presented in one evening, with a dinner at 5:00 p.m. and the program from 6:00 to 9:00 p.m. Classes are limited to no more than 30 participants.

Each class has participants from a number of different trades such as electrical, plumbing, sheet metal, dry wall, and painting. The classes encourage interaction among the different trades. This allows each participant to learn important aspects of the other trades and, at the same time, exposes them to the perceptions of the other trades of their area, all in a neutral environment. As a result, when the participants go out to jobs, they have a much better understanding of the trades and unique problems associated with other contractors with whom they must work. It also establishes a network so that the participants can call one of their colleagues from the class if a perplexing problem comes up on the job concerning that person's trade. Because of the success of the initial program, an advanced program is now offered as well.

The basic program consists of the following modules:

- Supervision on the construction site
- Developing effective communications and human relations skills
- Employee motivation
- Establishing a motivational climate on the jobsite
- Job scheduling and manpower planning
- Teamwork and coordination

The advanced program consists of the following modules:

- *Getting organized.* Planning, coordinating, and assessing jobsite productivity and what "getting organized" requires
- *Effective job controls.* Setting up check-offs to effectively control the jobsite; conducting effective meetings; short-order scheduling
- *Developing motivation power.* Motivating different personalities; delegating effectively; coaching workers for maximum performance

- *Advanced communications I.* Communicating effectively with different personality types; conflict resolution on the job
- *Advanced communications II.* The five-step method for solving people conflicts; building a positive working relationship with the boss; handling jobsite conflicts
- *Industry panel.* Roundtable discussion to learn about the roles, attitudes, conflicts, and concerns of developers, architects, and engineers.
- *Managing workforce diversity.* Understanding and managing a culturally diverse workforce; identifying stereotypes that can interfere with workplace productivity; developing a plan for managing workforce diversity

Employee Relations

Concept 9: Company-wide Employee Data Gathering

Companies need to know what is going on in the field from the point of view of field labor. This concept describes two approaches used to gather information from workers—workshop discussions and questionnaires—and summarizes some of the results to be expected from using these tools.

This concept incorporates a company-wide employee workshop program as a means of training, communication, motivation, and data gathering. Both employees and management benefit from this type of experience. One of the most effective types of training is a workshop that encourages each employee to participate. Not only do employees learn things that will enable them to perform better, but the workshop provides strong motivation for the employee and is a valuable source of information for the company.

The data gathering process is improved if more than one way is used to collect data. A second component of this concept incorporates questionnaires to confirm data collected in workshops, add detail to that information, and show trends over a period of time.

Two or three times a year, all employees in the company are brought into a workshop session to discuss the four key areas for a successful construction job: materials, tools, information, and manpower. Group size is limited to about 25 to 30 participants, so multiple sessions are needed. The sessions, lasting about four hours, are centered on discussion with considerable feedback from the employees about how they can work more effectively by improving the four key areas.

The workshops are followed up with questionnaires designed to probe for more details and more clearly define problems that surface during the workshops. The questionnaires are provided to field supervisory personnel and, over a period of time, can show trends in problem areas that tend to occur perennially.

Success in this concept requires a company culture supportive of mutual trust between employees and management. Employees must be open, honest and positive in their approach to discussing problems in the field. Success also requires that management places enough value on cultivating this open and trusting relationship so that they are willing to invest the time and money needed to support it. Finally, success requires that management follows through in dealing with problems when they surface. Nothing kills this type of program faster than when labor responds positively to management's request for participation by identifying problems and management ignores them.

The company presenting this concept has up to 150 union employees, including apprentices, at any given time. Each employee is brought into a four-hour workshop session, voluntarily, two or three times a year. The sessions are held immediately after the end of the workday (about 4:30 p.m.) and run until 8:30 p.m. or 9:00 p.m. A meal is provided. The sessions are held at the home office, which has facilities to accommodate about 40 to 50 people in a single group. The sessions are based upon discussion, with questions focusing on the four key areas identified above.

These sessions are a rich source of information on improvements that can be made at the worksite, from the people who are most knowledgeable about how to improve worker productivity. The workshops are followed by a survey sent to the field supervisors, who represent about 20% of the workforce. The purpose of the survey is to gain more of a consensus on critical items brought out during the workshop sessions. The survey is kept simple. A typical survey is composed of 16 questions that ask how the company is doing in one of the four key areas. The results of the survey are converted to percentages, which, over a period of time, show trends in company performance.

The data from both the workshops and the surveys provides the basis for feedback and continual improvement in critical company operations. It also provides a basis for measuring improvement. For example, in the area of tools, ratings early on were at a level of 16%. Over a 12-year period, these ratings have risen to 95%, demonstrating that the system works.

Acceptable ratings have been raised over the years. Some years back, only ratings of good, very good, or excellent were considered acceptable. Today, only ratings of very good or excellent are considered acceptable. Good is not good enough anymore!

Concept 10: Company Loyalty

Company loyalty is critical to productivity. A significant problem typically present when working with organized labor is where the loyalty lays: with the company or with the union. The problem of company loyalty is not just a "union" problem, nor is just a "blue-collar" problem. Studies and experience have both shown that one of the most important factors in determining levels of productivity is the relationship between employee and employer. Higher productivity is associated with stronger company loyalty, whereas lower productivity is associated with estrangement from the company.

Not only is loyalty important in productivity, but it is also important in terms of marketing. Field employees are the first line of contact with the customer. This is especially true with smaller companies and smaller jobs. Field employees must understand the importance of the company to them as employees if they are to be motivated to maintain high levels of productivity and if they are to represent the company well to the general public and to the company's market.

Loyalty in union companies is typically oriented initially toward the union, since apprenticeship training is where employees begin their careers. If the primary loyalty remains toward the union, rather than the employer, performance appears to suffer. To gain loyalty from field employees, one company has developed a program centered on frequent employee meetings. A minimum of six meetings are held each year to foster communication. The meetings are scheduled in such a way that one-half of the meeting is on company time, while the other half is on the employee's time. Employees are not required to remain for the second half of the meeting; however, they invariably chose to stay for the second half. This can be attributed to a number of things, including, but not limited to (1) a natural response to the efforts of the company, (2) interest in the content of the meeting, (3) opportunity to see friends and colleagues, and (4) refreshment provided at the end of the meeting.

Content of the programs encompasses a wide variety of topics, but one of the most important addresses the question of job security. Typical questions covered include what is the current status of the market? and how is the company pursuing additional work? It is important for employees to understand the true status of the market. If they do, and the result is a feeling of confidence in the security of their jobs, then they will work hard to help, not hinder job progress just to try to preserve a job in hard times. Meetings are scheduled to begin at 3:30 p.m. They are scheduled in the middle of the week so employees will not be inclined to stay out after the meeting rather than going home. When the meeting is over, the company provides free beverages and refreshments.

Information sharing is critical to maintaining loyalty. As much information as possible should be shared with all field employees. Some information is appropriate and relevant to all employees, while other information is appropriate only for foremen and supervisors. In order to implement this concept, care must be taken to focus on positive company matters and the complementary role of the union, rather than setting up an "us and them" environment. Care should also be taken to have a well-defined company drug and alcohol abuse policy that reinforces the idea of moderation in alcohol use, strictly off the jobsite. A typical meeting agenda enables discussion of topics such as (1) relevant safety items, (2) current factors affecting the industry, (3) general company information items, and (4) work status update, summarizing past, current, and anticipated work.

Concept 11: Performance-based Wage Scale

One major factor in being competitive is for a company to be able to produce higher-quality work at competitive rates. This requires higher performance levels from workers, who, in turn, deserve to be compensated at a higher level. With a flat wage rate, employees performing at a higher level are compensated at the same rate as those performing at lower levels. Though wage rate alone is not the only incentive or motivator for higher performance, recognition of superior performance in the form of a higher wage rate is an incentive to perform better. Wage rates linked to performance reward superior performance, but the effect is cyclical in that this recognition begets continued superior performance. Thus, compensation level tied to performance is a natural relationship and a clear productivity concept.

Labor agreements typically establish a minimum-wage level. Labor bargaining representatives, who represent the interest of the broad cross section of their local union, typically feel they best serve those interests by working to gain a rate as high as the market will bear. They also work to make this rate uniform so that no favoritism is shown toward their constituents. This environment is not conducive to a flexible rate based upon performance (or any other criteria). Even though the agreed-upon rate is a "minimum," it is typically so high that the employer has little or no room to create a flexible rate above it. Moreover, any differentiation between workers is strongly opposed.

For a performance-based wage rate to work, labor must recognize that if labor is rewarded commensurate with performance, individual performance will tend to improve, and those who perform poorly will be weeded out, creating a healthier, more productive labor pool. As a result, companies become more competitive and jobs become more plentiful while unemployment diminishes and market share is gained.

The minimum wage must be a valid minimum and set at a level low enough that receiving it is a disincentive. Labor must recognize the importance of rewarding workers for levels of performance: high performance justifiably gains high compensation and low performance justifiably gains low compensation. Management must develop fair and equitable evaluation procedures. Management must recognize that performance-based compensation is a key, but only one key to improve performance.

The company that provided this concept established a performance-based wage scale when there was high employment in the area because of the presence of a major project. This resulted in the attraction of a large labor pool from outside the area. Nonunion contractors were able to find ample supplies of labor willing to work at low wages. Union contractors were rapidly losing market share. This set up an environment that enabled labor to recognize that drastic measures were required, and they lowered the "minimum" rate to a competitive level and one that would allow room on top for performance rewards. As a result, market share was regained, even in a time of high competition. The key was labor working with management to apply innovative measures to restructure the wage scale. Management responded with a simple yet effective and fair rating process providing true rewards for superior performance.

This system continues today, and even though the minimum rate has increased since the performance scale was established, the scale has not changed. There has always been an adequate number of workers who perform at a level above even the higher minimum scale so that the contractor has had no difficulty in maintaining crews. Estimating work based upon this sliding scale has posed no problems. Company standard performance rates have been established and tied into a base level of compensation. Inevitably, when workers at a higher rate are used on a job, their performance is high enough to more than offset the higher wage rate.

A final benefit reported by the company was an increase in productivity. Not only has this productivity concept lowered costs, but production rates for work items have shown a 25% improvement. With significantly lower wage rates and significantly higher production rates, the company has been able not only to survive but to thrive in a hostile environment.

Company Management

Concept 12: Strategic Planning

Strategic planning is essential to any well-run company. It provides direction for company efforts and resources. It provides a basis for measurement against which performance can be evaluated. It provides valuable information that can be used to secure financing and bonding, as well as information useful in marketing. Because change is the norm—not just in the construction industry, but in all aspects of business life including business practices, economics, and technology—the strategic planning process must be repeated at regular intervals.

The company that contributed this concept carries out a major strategic planning every five years. The objective of the most recent planning cycle was to develop a strategy that would position the company as a leader in the use of technology.

A major strategic planning exercise usually requires the services of an outside facilitator to guide the company through the planning process. However, all the work (such as data gathering and analysis) is done by company personnel. The planning process for the company offering this concept took about seven months, with personnel spending many days sequestered away in meetings. Sessions focused on various topics. For example, one session was used to develop a mission statement. It took about nine hours. Nevertheless, the investment of time and effort to develop a well-thought-out mission statement is a critical part of the strategic planning process, since the mission statement forms the foundation upon which the strategic plan is based. Moreover, the process of developing a mission statement is invaluable to the participants. Another session involved the team in goal setting. They started by listing company goals, reaching 161 and ultimately boiling them down to only three.

In addition to giving the company direction and inspiration, a number of concrete products will result from the strategic planning process. One product emerging from this company's effort was a company policy manual. Another product was a well-designed financial plan that supports efforts to secure financing, insurance, and bonding. A third product was a card the size of a business card containing the corporate mission statement on one side and the three company goals on the other. This card is broadly disseminated throughout the company. Tangible information from the strategic planning process also forms a major component of the marketing brochure.

Top-level commitment is essential for successful implementation of a major strategic planning process. Part of this commitment must be to involve a broad cross section of the company in the process. This, in turn, requires a high level of trust in employees. Investment of both time and money is another significant component of the commitment. Before starting a strategic planning process, the contractor should seek out a professional to design a planning program commensurate with the size and complexity of the company. All companies need to have strategic plans. However, not all strategic plans must be as complex and extensive as the one described in this concept.

Concept 13: Project Manager Sales Analysis

Evaluating employee performance is always a difficult task. This can be especially difficult for project managers who often function independently on projects and do not have an immediate supervisor watching over their day-to-day performance.

One company has felt it important to be able to evaluate and monitor the performance of project managers. It has devised a means to track project manager performance, both on an ongoing basis and annually. This subsystem is integrated with other project management and accounting systems in the company.

The system consists of a series of six reports that highlight the current and cumulative performance of the project manager:

- *Sales analysis summary.* Provides a monthly and year-to-date sales analysis by project manager for both time and material sales and contract sales.
- *Work in progress summary.* Provides a monthly summary of work in progress by project manager for all contract jobs. The report breaks work down into estimated and actual amounts of materials, labor, and other categories. It also provides a compilation of percent complete and estimates of profit amount and profit percent. Finally, it gives a summary of billing status.
- *Work in progress comprehensive summary.* Provides a monthly summary of work in progress by project manager for all time and materials jobs. This report shows all project managers and has room to show the figures for several months at the same time.
- *Sales analysis detail.* Provides both time and materials sales and contract sales for a specific project manager on a monthly basis. The bottom line of this report constitutes a single line entry in the sales analysis summary.
- *Work in progress detail.* Provides work in progress information for contract jobs for a specific project manager on a monthly basis. The report includes a line for each job in progress for a given project manager during the month and provides information on estimated hours, actual hours, and percent complete.
- *Work in progress comprehensive detail.* Provides a monthly work in progress detail for a specific project manager for time and materials jobs. This report shows the detail behind the work in progress comprehensive summary.

This set of reports provides extensive information both on job status and on performance for each job and for each project manager. This system requires flexible project accounting and payroll systems to be in place so that the family of reports can be developed to provide the selected information.

Concept 14: Manpower Projections

To work effectively, companies must maintain a level usage of labor. That is, it is important to keep good people working and minimize the number of people hired for a short period of time. Giving employees the opportunity to have permanent employment makes them feel part of the company. Minimizing short-time hires reduces costs associated with mobilizing new workers and minimizes risk in acquiring workers below the company standard.

Manpower leveling has often been used in large construction jobs to avoid short-term peaks and valleys where production suffers either from lack of personnel (at peak times) or from idle workers (at slow times). This concept extends the idea of manpower leveling to the company level by targeting jobs to bid or negotiate that will support the objective of maintaining a level worker resource. The market must be strong enough that there is opportunity to pass up some available jobs if worker leveling is desired.

The company that contributed this concept has developed a simple spreadsheet to provide workday projections for company labor usage. The sheet shows months as columns and different jobs as line items, with two lines for each job. The first line is a projection of monthly labor budget in terms of dollars. The second line converts the monthly dollar projection to a workday projection. At the bottom of each column (month), both total labor cost budgets and total worker-day projections for the given month are provided. Annual projections of both labor costs and workdays are given for a given job in a "totals" column at the right of the sheet. The summation at the bottom of the right-hand "totals" column provides a projection of labor dollars and workdays for the year. One additional line has been added to the monthly totals across the bottom of the sheet that converts workdays projected for the month to workers required for the month.

This spreadsheet can show at a glance when peak workloads or slack workloads are anticipated. The estimators then know when to hold pricing particularly tight, or when they need to add in a bit more contingency for staffing a job with noncompany workers. If the company has a person responsible for marketing, it can use this information to focus on those jobs occurring during projected slack times, and to move away from jobs occurring during peak work usage times.

This concept has a number of other uses as well. It will support macro planning for labor usage across the company so that project managers or superintendents know which jobs to call when they need additional workers. It can also be used to provide valuable data for labor planning when delays are anticipated on a given project. Finally, it can be used to support claims when work progress has deviated from the schedule provided either at bid time or by the general contractor, or in the event the specialty contractor's work has been delayed by another contractor.

This concept requires information from estimators and/or project managers on projected labor costs and worker needs, by month, for each job under contract. Setting up the spreadsheet requires only a basic understanding of spreadsheet fundamentals. The spreadsheet should be updated every time a project is placed under contract and whenever changes affecting worker usage (either amount or timing) are made to jobs under contract. Reports should be produced on a monthly basis and distributed to all project managers, superintendents, and company managers.

Concept 15: Backlog and Bidding Report

Marketing is an important component of a construction company. Part of the marketing function is to be aware of the backlog of work the company has at any given time. Backlog comprises several aspects, including volume of work to be completed under contract, anticipated profit, and "survival time" which indicates how many months the company can continue with no additional work before the overhead eats up the profit for the year. This concept is appropriate for any company, independent of size, geographic location, or the state of the economy. It relates a good business practice that will provide contractors with essential information on the current status of their business, as well as a valuable planning tool.

The company that contributed this concept prepares a report with five columns entitled Backlog, Projected Backup Rate, Amount Earned, to Bill Fiscal Year, and Expected Billings for Fiscal Year. Line items represent months 1 to 12 of the current fiscal year. Data is entered monthly on the line corresponding to that month. The amount in the Backlog column will fluctuate depending on what new jobs have been placed under contract during the previous month and what has been billed out. Projected Markup Rate will show a small variation around the base rate at which the company is operating. Should this significantly change, it will flag a serious change in the way the company is operating. Amount Earned will increase throughout the year as payments are made on work in progress. To Bill Fiscal Year will increase and decrease as backlog varies, but will move toward zero as the fiscal year closes and new backlog is pushed into the next fiscal year. Expected Billings for Fiscal Year is simply the sum of Amount Earned and To Bill Fiscal Year. It will increase as new work is added to the backlog, but at a decreasing rate as new backlog is pushed into the new fiscal year. It will converge to the total billings for the fiscal year at the end of the last month.

Anticipated total fiscal year gross profits are calculated as the sum of gross profits on earnings and expected gross profit. Survival time is calculated by dividing expected gross profits by monthly overhead.

This concept can be implemented with information any company accounting system should have readily available on the status of work under contract, plus information from the estimating department on markups. Development of the application on the spreadsheet requires only a basic level of spreadsheet literacy. Reports should be updated on a monthly basis and used for company level planning. Therefore, reports should be provided to upper-level management only.

CONCLUSIONS

In addition to the productivity concepts gathered through the focus group workshops, a number of useful general observations emerged from the study. The most important observation is that successful contractors share certain characteristics, and these characteristics were common to virtually all the contractors who actively participated in the focus group workshops. Contractors may have different approaches to how they accomplish tasks at the detail level, but overall, certain attitudes and general approaches are always present in the successful contractors:

- *Flexibility in changing times.* Participants agreed that there is great change in the construction industry. They agreed further that this is not a phase, but change has become a constant in the industry. Change should not be a curious thing to a contractor because it is a rare project indeed that is completed without changes in the scope of work. Recognizing that change is a normal part of the construction environment is key to being a successful contractor. Some of the areas that are changing include technology, management approaches, the economy, and the market itself. Contractors must be flexible in changing times. They must view change not as a threat but rather as an opportunity. Change will enable the successful contractor to become more successful and to outstrip competitors who refuse to recognize that change is a fact of life in the industry today. As a corollary to the recognition of change, successful contractors are never satisfied. They are continually searching for a better way to do things and adjusting operations to make them better. They are measuring the change to enable tracking it, controlling it, and shaping it in order to form a better company.
- *Comfort in the workplace.* A strong feeling of comfort issued from the successful contractors. Even though change is constant, they feel in control. Even though problems crop up on jobs, they know that they can solve the problems and successfully complete jobs. Even though there is great competition and pressure on market share, they feel confident they can remain competitive and expand into new markets as they emerge. Workers at each level also feel comfortable. Labor feels that they are a part of the company, not estranged from or in conflict with it. Supervisors feel in control. They feel they are working with labor and are supported by management. Project managers express trust, both toward those who are working for them and those for whom they work. They feel confident that labor is skilled and super-

vision is competent, and they know that upper-level management provides all the support needed to successfully complete their projects. Finally, top-level managers express confidence in all their employees from top to bottom throughout the company. Top-level management also knows their employees and stays in touch with them. This is what gives them the confidence.

- *Project-level control.* All participants agreed that project control is essential. Well-defined procedures must be executed and monitored constantly to maintain control. Estimating and bidding, planning, mobilization, and labor tracking were all identified in productivity concepts as keys to successful company operations.
- *Responsibility and accountability from the top down.* Successful companies thrive on responsibility and accountability at each level. These emanate from the highest level in the company and permeate to the lowest levels. Top-level management keeps in touch with all levels of employees. They establish contact, disseminate information, listen to comments, and respond to each communication. Most importantly, they express care for each individual in the company. This caring attitude is also reflected outside the company. The successful company demonstrates real care for all with whom it comes in contact.
- *Strategic planning.* Strategic planning is important at all times but becomes critical in times of change. Successful contractors understand their business, understand the environment surrounding the business, and adapt the business to the environment through the strategic plan.

APPENDIX A
BIBLIOGRAPHY

2005 Baldrige National Quality Program. (2005). *Criteria for Performance Excellence.* United States Department of Commerce, National Institute of Standards and Technology, Gaithersburg, MD.

Anderson, T. (1992). "Step into My Parlor: A Survey of Strategies and Techniques for Effective Negotiation." *Business Horizons,* 35 (3): 71–76.

American Society for Quality. (2000). Quality Management Systems: Fundamentals and Vocabulary. ANSI/ISO/ASQ Q9000-2000.

———. (2000). *Quality Management Systems: Guidelines for Performance Improvements.* ANSI/ISO/ASQ Q9004-2000.

———. (2000). *Quality Management Systems: Requirements.* ANSI/ISO/ASQ Q9001-2000.

American Quality Foundation and Ernst & Young. (1992). *Best Practices Report: An Analysis of Management Practices That Impact Performance.* New York: AQF/Ernst & Young.

Bogan, C. E., and English, M. J. (1994). *Benchmarking for Best Practices.* New York: McGraw-Hill.

———. (1980). *Scheduled Overtime Effects on Construction Projects, Report C-2.* Construction Industry Cost Effectiveness Project, New York.

The Business Roundtable. (1982). *Absenteeism and Turnover.* Report C-6, New York.

Camp, R. C. (1989). *Benchmarking: The Search for Industry Best Practices That Lead to Superior Performance.* Milwaukee, WI: ASQC Quality Press.

——. (1994). *Business Process Benchmarking: Finding and Implementing Best Practices.* Milwaukee, WI: ASQC Quality Press.

Case, J. (1994). "Games Companies Play." Inc.com: http://www.inc.com/magazine/19941001/3137.html. Accessed on April 5, 2008.

Cohen, A. (2000). *The Relationship between Commitment Forms and Work Outcomes: A Comparison of Three Models. Human Relations* 53 (3): 387–417.

Dawson, V. (1999). *Lincoln Electric: A History.* Cleveland: Lincoln Electric Company.

Department of Trade and Industry. (1995). *Best Practice Benchmarking.* London.

Fischbach, A., and Harrington, M. (2002). "Top 50 Electrical Contractors Face Tough Times." *CEE News* 54 (6): 10.

Galloway, D. (1994). *Mapping Work Processes.* Milwaukee, WI: ASQC Quality Press.

Glavinich, T. E. (1995). *Quality Assurance Guide for Inside and Outside Electrical Contractors.* Bethesda, MD: The Electrical Contracting Foundation, Inc.

Glavinich, T. E., and Rowings, J. E. (1994). *Total Quality Management for the Electrical Contracting Industry.* Bethesda, MD: The Electrical Contracting Foundation, Inc.

Lincoln, S., and Price, A. (1996). "What Benchmarking Books Don't Tell You." *Quality Progress,* March, 33–36.

Hammer, M. (1996). *Beyond Reengineering.* New York: HarperBusiness.

Hammer, M., and Champy, J. (1993). *Reengineering the Corporation.* New York: HarperBusiness.

Harrington, H. J. (1991). *Business Process Improvement: The Breakthrough Strategy for Total Quality, Productivity, and Competitiveness.* New York: McGraw-Hill.

Hessen, C. (2000). "Using an Incentive Compensation Plan to Achieve Your Firm's Goals." *Journal of Management in Engineering* 16 (1), 31–33.

Hinze, J. (1985). "Absenteeism in the Construction Industry." *Journal of Management in Engineering,* ASCE 1 (4): 188–200.

Huff, A. (2001). "Tempting Drivers to Excel." *Commercial Carrier Journal* 158 (12): 22–26.

Karasek, R. A., and Theorell, T. (1990). *Healthy Work: Stress, Productivity, and the Reconstruction of Working Life.* New York: Basic Books, Inc.

Kelley, S. (2002). "Sharing the Wealth." eTracker.com: http://www.etrucker.com/apps/news/article.asp?id=16006. Accessed on April 5, 2008.

Leibfried, K. H. L., and McNair, C. J. (1992). *Benchmaking: A Tool for Continuous Improvement*. New York: HarperBusiness.

Lincoln, J. F. (1946). *Lincoln's Incentive System*. New York: McGraw-Hill.

McDonald, P., and Baldwin, G. (1989). *Builder's and Contractor's Handbook of Construction Claims*. Englewood Cliffs, NJ: Prentice Hall.

McKenzie, J. (2000). "The 2000 Profile of the Electrical Contractor." *Electrical Contractor* 65 (6), 36–40.

McKenzie, J., and Bailey, D. (2001). "Guide to the Electrical Contracting Market." Electrical Contractor.com. http://www.ecmag.com/research/guide/2001guide.pdf. Accessed on March 6, 2001.

Mechanical Contractors Association of America. (1968). *Tables for Calculation of Premium Time and Inefficiency on Overtime Work*. Bulletin 20, Washington, D.C.

Mobley, W. M., Horner, S. O., and Hollingsworth, A.T. (1978). "An Evaluation of Precursors of Hospital Employee Turnover." *Journal of Applied Psychology* 63 (4), 408–414.

———. (1983). *Rate of Manpower Consumption*. Index No. 5075, Bethesda, MD.

———. (1989). *Overtime and Productivity in Electrical Construction*. Index No. 5050, Bethesda, MD.

National Electrical Contractors Association. (2000). *Electrical Contractors' Financial Performance Report*, Bethesda, MD.

Nicholas, J. (2001). *Driver Retention, Character Move to the Forefront*. E-Tracker: http://www.etrucker.com/apps/news/article.asp?id=1147. Accessed on April 5, 2008.

Parker, G., McAdams, J., and Zielinski, D. (2000). *Rewarding Teams: Lessons from the Trenches*. San Francisco: Pfeiffer & Company.

Peck, C. (1995). *Individual Incentive Programs*. New York: Conference Board, Report No. R-1127.

Rolstadas, A. (1995). *Benchmarking: Theory and Practice*. London: Chapman and Hall.

Schnake, M., and Dumler, M. P. (2000). *Predictors of Propensity to Turnover in the Construction Industry. Psychological Reports* 86: 000–1002.

Sall, J., Lehman, A., and Creighton, L. (2000). *JMP Start Statistics*. 2nd ed. Pacific Grove, CA: Duxbury.

Sibson, R. E. (1990). Compensation. 5th ed. New York: AMACOM.

Smith, A. (1987). *Increasing Onsite Productivity.* Transactions, AACE, Morgantown, WV, K.2.1-K.2.10.

Spendolini, M. J. (1992). *The Benchmarking Book.* New York: American Management Association.

Thomas, H. R. (1994). "Construction Productivity Losses from Overmanning: A Literature Review and Analysis." Department of Civil Engineering, The Pennsylvania State University, University Park, PA.

Thomas, H. R., and Oloufa, A. (1996). *Labor Inefficiencies Caused by Schedule Acceleration and Compression.* Report to the Electrical Contracting Foundation, Inc., Bethesda, MD.

Thomas, H. R., Oloufa, A., Hanna, A., and Noyce, D. (1995). *What-To-Do-Guide for Schedule Acceleration and Compression.* Index No F9503, The Electrical Contracting Foundation, Inc., Bethesda, MD.

Thomas, H. R., and Raynar, K. (1994). *Effects of Scheduled Overtime on Labor Productivity: A Quantitative Analysis.* Report to the Construction Industry Institute, Austin, TX.

Thomas, H. R., and Smith, G. (1990). *Loss of Construction Labor Productivity due to Inefficiencies and Disruptions: The Weight of Expert Opinion.* PTI Report 9019, Pennsylvania Transportation Institute, University Park, PA.

U.S. Army Corps of Engineers. (1979). *Modification Impact Evaluation Guide.* EP 415-1-3.

Watson, G. H. (1993). *Strategic Benchmarking: How to Rate Your Company's Performance Against the World's Best.* New York: John Wiley & Sons.

Womack, J. P., and Jones, D. T. (1996). *Lean Thinking.* New York: Simon and Schuster.

Womack, J. P., Jones, D. T. and Roos, D. (1991). *The Machine That Changed the World.* New York: HarperCollins.

Zairi, M. (1994). *Measuring Performance for Business Results.* London: Chapman and Hall.

APPENDIX B
FORMULAS

LABOR PRODUCTIVITY FORMULAS

Company-Level Labor Productivity Formulas

$$P_{COMPANY\text{-}CURRENT} = \frac{TAVA}{TAL} \quad (1.1)$$

where:

$P_{COMPANY\text{-}CURRENT}$:	Productivity at the company level (not adjusted for wage inflation)
TAVA:	Total accrued value added
TAL:	Total accrued labor hours

TAVA is calculated as:

TAVA = Acrrued [Income–Material–Subcontacts–Owner paid overtime premiums (1.2)

$$P_{COMPANY} = \left(\frac{TAVA}{TAL}\right)\left(\frac{WBY}{WCY}\right) \quad (1.3)$$

where:

$P_{COMPANY}$: Productivity at the company level

TAVA:	Total accrued value added
TAL:	Total accrued labor hours
WBY:	Average wages base year
WCY:	Average wages current year

$$P_{COMPANY\text{-}HOURS} = \left(\frac{TAVA}{TAL}\right)\left(\frac{WCY}{WBY}\right)(VAM) \qquad (1.4)$$

where:

$P_{COMPANY\text{-}HOURS}$:	Productivity at the company level (in hours per given revenue)
TAL:	Total accrued labor hours
TAVA:	Total accrued value added
WCY:	Average wages current year
WBY:	Average wages base year
VAM:	Value-added multiplier

$$P\text{-}Index_{COMPANY} = \left(\frac{TAVA}{TAL}\right)\left(\frac{WBY}{WCY}\right)(IF) \qquad (1.5)$$

where:

$P\text{-}Index_{COMPANY}$:	Productivity index at the company level
TAVA:	Total accrued value added
TAL:	Total accrued labor hours
WBY:	Average wages base year
WCY:	Average wages current year
IF:	Index factor

The value of IF can be calculated by using a simple formula:

$$IF = \frac{100}{P_{COMPANY} \text{ for the base year}} \qquad (1.6)$$

Field-Level Labor Productivity Formulas

$$P_{FIELD} = \left(\frac{TAVA}{AFL}\right)\left(\frac{WBY}{WCY}\right) \qquad (1.7)$$

where:

P_{FIELD}:	Productivity at the field level
TAVA:	Total accrued value added
AFL:	Accrued field labor hours
WBY:	Average wages base year
WCY:	Average wages current year

$$P_{FIELD\text{-}HOURS} = \left(\frac{AFL}{TAVA}\right)\left(\frac{WCY}{WBY}\right)(VAM) \tag{1.8}$$

where:

$P_{FIELD\text{-}HOURS}$:	Productivity at the field level (in hours per given revenue)
AFL:	Accrued field labor hours
TAVA:	Total accrued value added
WCY:	Average wages current year
WBY:	Average wages base year
VAM:	Value-added multiplier

$$P\text{-}Index_{FIELD} = \left(\frac{TAVA}{AFL}\right)\left(\frac{WBY}{WCY}\right)(IF) \tag{1.9}$$

where:

$P\text{-}Index_{FIELD}$:	Productivity index at the field level
TAVA:	Total accrued value added
AFL:	Accrued field labor hours
WBY:	Average wages base year
WCY:	Average wages current year
IF:	Index factor

Project-Level Labor Productivity Formulas

$$P_{PROJECT} = \left(\frac{TAVAP}{ALH}\right)\left(\frac{WBY}{WCY}\right) \tag{1.10}$$

where:

$P_{PROJECT}$:	Productivity at the project level
TAVAP:	Total accrued value added of project
ALH:	Accrued labor hours for the project
WBY:	Average wages base year

WCY: Average wages current year

$$\text{P-Index}_{\text{PROJECT}} = \left(\frac{\text{TAVAP}}{\text{ALH}}\right)\left(\frac{\text{WBY}}{\text{WCY}}\right)(\text{IF}) \quad (1.11)$$

where:

P-Index $_{\text{PROJECT}}$:	Productivity index at the project level
TAVAP:	Total accrued value added of project
ALH:	Accrued labor hours for the project
WBY:	Average wages base year
WCY:	Average wages current year
IF:	Index factor

Activity-Level Labor Productivity Formula

$$\text{P}_{\text{ACTIVITY}} = \left(\frac{\text{TAVAA}}{\text{ALHA}}\right)\left(\frac{\text{WBY}}{\text{WCY}}\right) \quad (1.12)$$

where:

$\text{P}_{\text{ACTIVITY}}$:	Productivity at the activity level
TAVAA:	Total accrued value added of the activity
ALHA:	Accrued labor hours for the activity
WBY:	Average wages base year
WCY:	Average wages current year

STACKING OF TRADES FORMULAS

Productivity Efficiency Formula

$$P = 3.288 + 1.67(A) - 3.173(B) + 0.4962(B^2) - 0.3225[(A)(B)] \quad (3.1)$$

where:

$$P = Ln\left(\frac{\text{Actual hrs}}{\text{Estimated hrs}}\right)$$

$A = Ln$ (Electrical density in ft2/worker)
$B = Ln$ (Nonelectrical density in ft2/worker)
Ln = Natural logarithm

Productivity Efficiency for a Total Density Model Formula

$$P = -4.642 + 2.108(C) - 0.2161(C^2) \tag{3.2}$$

where:

$$P = Ln\left(\frac{\text{Actual hrs}}{\text{Estimated hrs}}\right)$$

$C = Ln$ (Total density in ft2/worker)
Ln = Natural logarithm

Simplified Procedure Formula

$$A = 3.29 + \text{Equation value 1} + \text{Equation value 2} + \text{Equation value 3} \tag{3.3}$$

SCHEDULE ACCELERATION FORMULAS

Actual Weekly Labor Consumption Rate Formula

$$\text{Weekly labor consumption rate} = \left(\frac{\text{Weekly craft work hours}}{\text{Total craft work hours}}\right)(100) \tag{4.1}$$

Labor Rate Deviation Formula

$$\text{Labor rate deviation} = \text{Actual consumption rate} - \text{Planned consumption rate} \tag{4.2}$$

Gross Inefficient Work Hours Formula

$$\text{Gross inefficient work hours} = \text{Actual work hours} - \frac{\text{Actual work hours}}{\text{Performance ratio (PR)}} \tag{4.3}$$

Weekly Loss of Efficiency Formula

$$\text{Weekly loss of efficiency (\%)} = \left(1.0 - \frac{1.0}{\text{Weekly performance ratio}}\right)(100) \tag{4.4}$$

Project Loss of Efficiency Formula

$$\text{Project loss of efficiency (\%)} = \left(\frac{\text{Inefficient hours}}{\text{Actual hours} - \text{Inefficient hours}}\right)(100) \tag{4.5}$$

TURNOVER AND ABSENTEEISM FORMULAS

Absenteeism Formula

$$PL = -0.716 + 0.876(A) - 0.198(A)^2 \tag{6.1}$$

where:

PL = Productivity loss
A = Percent absenteeism

Turnover Formula

$$PL = -0.396 + 0.603(T) - 0.147(T)^2 \tag{6.2}$$

where:

PL = Productivity loss
T = Percent turnover

FIELD INCENTIVE SYSTEMS FORMULAS

Profitability Formula

$$\text{Points} = \left(\frac{\text{Total project profit}}{\text{Planned project profit}}\right)\left(\frac{\text{Hours assigned to project}}{\text{Total working hours in plan cycle}}\right) \tag{8.1}$$

Schedule Formula

$$\text{Points} = \left(\frac{\text{Planned working days}}{\text{Actual working days}}\right)\left(\frac{\text{Hours assigned to project}}{\text{Total working hours in plan cycle}}\right) \tag{8.2}$$

INDEX

B

C

Q

T